机电专业英语项目化教程
(微课版)

周　苑　窦爱女　主　编

清华大学出版社
北　京

内 容 简 介

"专业英语"是机电一体化技术专业的重要课程,为了建设好此课程,课程建设团队着手编写了本教材。本教材以机电行业为背景,主要介绍现代电气应用技术和先进制造技术两大部分。

本书以项目构建教学体系,使用工作页教学模式,选取企业中真实案例作为背景,选取当前最新的技术文献资料。在完成每一个任务的过程中进行新知识、新技能的学习和训练,将知识的系统性、完整性与实用性相结合。内容覆盖了机电专业大部分的学科专业知识,并补充了现代全新的技术应用,拓展了学生的视野。同时有配套的课件、相关多媒体音频、视频资料和工作页答案帮助授课教师在讲授过程中更好地运用该教材。本书可作为本科、高职高专院校机电相关专业的专业英语教材,也可供工程技术人员参考。

图书在版编目(CIP)数据

机电专业英语项目化教程:微课版/周苑,窦爱女主编. —北京:清华大学出版社,2022.5(2025.2重印)
ISBN 978-7-302- 57425-5

Ⅰ. ①机… Ⅱ. ①周… ②窦… Ⅲ. ①机电工程—英语—高等职业教育—教材 Ⅳ. ①TH

中国版本图书馆 CIP 数据核字(2021)第 019666 号

责任编辑:梁媛媛
装帧设计:李 坤
责任校对:周剑云
责任印制:沈 露

出版发行:清华大学出版社
　　　　网　　址:https://www.tup.com.cn, https://www.wqxuetang.com
　　　　地　　址:北京清华大学学研大厦 A 座　　邮　　编:100084
　　　　社 总 机:010-83470000　　　　　　邮　　购:010-62786544
　　　　投稿与读者服务:010-62776969, c-service@tup.tsinghua.edu.cn
　　　　质量反馈:010-62772015, zhiliang@tup.tsinghua.edu.cn
　　　　课件下载:https://www.tup.com.cn, 010-62791865
印 刷 者:涿州市般润文化传播有限公司
经　　销:全国新华书店
开　　本:185mm×260mm　　印　张:8.5　　字　数:205 千字
版　　次:2022 年 5 月第 1 版　　　印　次:2025 年 2 月第 4 次印刷
定　　价:29.00 元

产品编号:072591-02

前　言

随着世界加工制造业转移和工业化进程的加快，中国已经成为"世界制造中心"。传统的机械制造行业在经历了设备改造、技术革新之后出现了前所未有的发展势头，国外高新技术、新工艺迅速涌入国内，机械行业的加工手段越来越多，加工精度、自动化程度也越来越高，对人才也提出了更高的要求。为了满足企业要求，各高校很重视对学生专业英语能力的培养。

"专业英语"是机电一体化技术专业的重要课程，为了建设好此课程，课程建设团队着手编写了此教材。本教材以机电行业为背景，主要介绍现代电气应用技术和先进制造技术两大部分。

本教材在编写过程中，遵从淡化理论、够用为度、培养技能、重在实用的原则；在内容选择上，基于学生的认知规律，从简单过渡到复杂，从单一过渡到综合，从低级过渡到高级，并结合机电专业开设的专业课作为教学的载体，加强专业英语职业能力培养的针对性和实用性。

为了方便专业英语教师的教学和学生的学习，本教材主要采用了工作页模式，选取的都是企业真实案例中使用到的技术文献。在教学方法上以任务驱动组织教学。在完成每一个任务的过程中进行新知识、新技能的学习和训练，将知识的系统性、完整性与实用性相结合。内容覆盖了机电专业大部分的学科专业知识，并补充了现代全新的技术应用，拓宽了学生的视野。同时有配套的课件和工作页答案帮助授课教师在讲授过程中更好地运用本教材。本教材可作为本科、高职高专院校机电相关专业的专业英语教材，也可供工程技术人员参考。

本教材第一部分基础知识由上海电子信息职业技术学院机械与能源工程学院周苑编写；第二部分现代电气应用技术由上海电子信息职业技术学院机械与能源工程学院周苑编写，上海电子信息职业技术学院中德工程学院李康参与编写；第三部分先进制造技术由上海电子信息职业技术学院机械与能源工程学院窦爱女编写，傅卫沁参与编写。本书课件由上海电子信息职业技术学院机械工程系郭瑜心参与制作。本书在编写的过程中也得到了上海电子信息职业技术学院各位领导、各位教师和相关企业专家的大力支持，并参考了大量的文献资料，在此一并表示感谢。

由于编者水平有限，书中错误与不当之处在所难免，敬请读者批评指正。

编　者

Contents

Module 1 Basic Knowledge

1.1 Introduce Myself

Task Content: How to introduce myself? How to introduce my responsibilities? How to begin my career in workshops? Let's begin! Catch our step!

 Introduce Myself and My Company

Use the words and expressions in the word box to write a few sentences about you and your company.

And usually, the introduction will include the following topics.

- The name of your company
- The location
- The number of employees
- The production of your company
- The type of customers
- Your trade

It's called...	located in...	manufactures....
exports....	repairs....	joint ventures
private enterprises	electrical components	metalworking
fitter	technician	engineer

李军供职于一家 PLC 贸易销售公司，请为他写一篇自我简介。

大家好，我叫李军，我是一名电气工程师，现供职于一家 PLC 贸易销售公司。我公司主要负责 X 品牌的 PLC 的销售和售后支持。公司现位于上海市浦东新区。公司总人数为 59 人。我主要负责食品包装类 PLC 产品的整体销售和售后。我有 3 年从业经验，并有全自动食品包装机的大量项目实例经验。如果你有相关方面的需求，欢迎随时和我联系。

Dialogue

Micheal (The manager) and Li Jun are talking about the production plan. Please read the dialogue.

Micheal: Li Jun, how is everything going?

Li Jun: I have just finished the production plan for next month. But it still has some problems. I need your help.

Micheal: OK. What is the problem?

Li Jun: Mr. Muller applied vacation for next month. And we have the order which must use CNC to manufacture the curve. But Muller will go. I am now thinking about another way to do this curve.

Micheal: How about Mark? I know he is familiar with CNC. Although he hasn't operated CNC for many years, he had operated CNC in his previous company. Maybe he can do this.

Li Jun: Yes, I know that. But Mark now is busy with measuring. He is the only one who can use the coordinate measuring machines. He is the only technician!

Micheal: OK. I got it. How about using wire-electrode cutting? Linda can do it.

Li Jun: But Linda's schedule is full. That means she must work overtime.

Micheal: I know this. I will talk to Linda about overtime. And I will call you later.

Li Jun: OK. See you later.

 Extend Knowledge

Every workshop has a special trade, called a scheduler. A scheduler is also called a production planner. He/She is in charge of processing technology and planning every process task time. Usually, there are two groups of schedulers in one workshop. One group is in charge of plan processing step, the other is in charge of task time. If you become an operator in workshop, then you will team up with the schedulers day by day. And they will plan your vacation and your shift. In preceding part of the text, Li Jun is a scheduler.

Please write down the Chinese meanings of new words or phrases in the word box.

curve	CNC operator	previous company
_____	_____	_____
coordinate measuring	machine	technician
_____	_____	_____
wire-electrode cutting	schedule	overtime
_____	_____	_____

Reading Comprehension

Job Code: CNC20150325-1

Immigration Opportunity: Yes

Position: CNC Programmers

Pay Rate: 35 000~50 000 US dollars per year

Job Description:

Reads and interprets blueprints.

Properly and safely loads and unloads castings and raw materials in CNC machinery.

Uses gags and measuring instruments to meet tolerance requirements.

Fully understand the process and the technology of CNC program.

Position: CNC Operator

Pay Rate: 14~18 US dollars per hour

Job Description:

Reads and interprets blueprints.

Being acquainted with the turning, milling processing, properly and safely use the measuring implements, cutters, fixtures.

Being acquainted with the material of parts, dimensional tolerance, form and location tolerance, surface roughness and other technical requirements.

Here are two candidates of these two jobs. Please match them.

I am Allen Wang. I have the bachelor's degree of machine manufacturing. I have worked in a multinational corporation for 5 years. I am very familiar with CNC manufacturing and CNC programmer. And I know the processing technology. I can analyze the blueprints and plan the processing. Also I have good commands of English reading, writing and speaking.

My name is May Lin. I am 21 years old. I have just graduated from vocational school. My major is mechanical-electrical integration. I am now searching for a job. I prefer CNC operator. I have no work experience. I need to learn!

Writing

Please write something about yourself and talk with your partner.

 请和教师交流结果。

教师签名：_____

1.2　My Workshop

Task Content: Peter Karl comes from United Kingdom. He is a new worker in your company. Please introduce your workshop and your workshop safety rules to him.

Here are some safety signs. Please choose which can be seen in the workshop.

(1)

(2)

(3)

(4)

(5)

(6)　　　　　　　　　(7)　　　　　　　　　(8)

(9)　　　　　　　　　(10)

上述标识中有没有不属于车间的标识?

你知道这些标识都是什么意思吗?请在群组内讨论,这些标识可能出现在什么设备上。

你还见过什么样的警告标识?

Dialogue

Peter is your new colleague. He is now reading the safety rules of the workshop. Please explain the rules to him.

Peter: Hi, my name is Peter Karl. I am new here. Nice to meet you.

You: Nice to meet you. I have worked here for 3 years.

Peter: Tell me something about this workshop's safety rules, please.

You: OK. First you must wear safety equipment, including overalls, work shoes, protective goggles. When you use the jigsaw, you must wear ear protectors to avoid the huge noise. And when you use the drilling machine, you can't wear gloves. Because wearing gloves will cause more serious accident.

Peter: Yes, I know that. I have worked in another metalwork workshop before. I have been a fitter.

You: That's good! I'm sure that you will soon like here!

Peter: Thank you. I hope so. And have lunch together?

You: Good, my pleasure. This way, please.

请在群组内讨论，还需要补充什么安全注意事项，可以使用以下句型。

Please don't..., when...

Remember do/don't...

If you want to..., don't forget...

If you don't ..., you will...

It is forbidden that...

You can..., but remember...

The... are stored in ...

... is next to the ...

Please write down the Chinese meanings of new words or phrases in the word box.

workshop	equipment	overall	
_____	_____	_____	
work-shoe	protective goggle	jigsaw	fitter
_____	_____	_____	_____
drilling machine	protective glove	metalwork	
_____	_____	_____	

Reading Comprehension

The following procedures will help prevent injuries and increase efficiency.

- Keep all tools and service equipment in good condition.
- Always use the appropriate personal protective equipment for operations such as welding and grinding.
- Keep floors and benches clean to reduce fire and tripping hazards.
- Clean the area completely after a job is finished.
- Empty trash containers regularly. Never store oily, greasy rags in closed containers—this practice has been responsible for numerous fires due to spontaneous combustion.
- Lighting, wiring, heating, and ventilation systems should be properly maintained.
- Do not allow unauthorized use of tools, service equipment and supplies.
- Don't allow anyone to use tools or service equipment without proper instruction.
- Keep guards and safety devices on power tools in place and functional.
- Use tools and service equipment only for their designed purposes.
- Service fire extinguishers on a regular schedule.
- Keep the first aid kit fully stocked.

Read the text, and decide whether each of the following statements is True(T) or False(F).

1. I needn't clean my workplace everyday.   □T   □F
2. After a job, I must put the tools into storage cabinets.   □T   □F
3. I need an electric certificate when I connect motor circuit.   □T   □F
4. Fire safety checking is the task of a firefighter.   □T   □F
5. I can lend tools to my friend who use it to fix his roof.   □T   □F

和你的同伴讨论一下，你得到答案的依据是上述条例中的哪一条。

Translate the following sentences into Chinese.

1. Do not allow unauthorized use of tools, service equipment and supplies.

_____

2. Use tools and service equipment only for their designed purposes.

_____

3. Clean the area completely after a job is finished.

_____

4. Always use the appropriate personal protective equipment for operations such as welding and grinding.

_____

5. Don't allow anyone to use tools or service equipment without proper instruction.

_____

Writing

Please write the safety rules for the electric lab.

The following procedures will help prevent injuries and increase efficiency.
- *...must be worn, when you go into the lab.*
- *Keep away from... if you don't know the function of that.*
- *Keep...clean to reduce fire and tripping hazards.*
- *Keep...clean after you finish your job.*
- *Wiring and ...should be properly placed in...*
- *Do not allow ... use of tools, service equipment and supplies.*
- *Don't allow ... to closed the electric loop without...*
- *Don't bring ... into the electric lab.*
- *Use...only for their...*
- *...and...is forbidden in this lab.*
- *Ensure the fire extinguishers are ...*
- *When you are in accident condition, please ... and don't...*

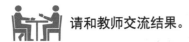 请和教师交流结果。

教师签名：＿＿＿＿＿＿＿

1.3　Trades & Tools

Task Content: How many trades are there in a workshop? What is the difference of these trade? Which trade do you prefer? How to use the tools? How to call all these tools?

 Trades

Please translate these words or phrases in the word box into Chinese.

fitter	CNC operator	lathe operator
＿＿＿＿＿＿	＿＿＿＿＿＿	＿＿＿＿＿＿
customer service	miller	programmer
＿＿＿＿＿＿	＿＿＿＿＿＿	＿＿＿＿＿＿
technician	engineer	
＿＿＿＿＿＿	＿＿＿＿＿＿	

E-mail from May Lin

Dear Anna,

    Hello, Anna. Is everything going well?

    Now I have stayed in Shanghai for about two weeks. You can't imagine that I found a job here. Now I have worked in this company for a week. The new job is really a great challenge for me. It is an automation equipment manufacture company. My responsibility is quality control. And I haven't done this kind of job before! I had learned how to use a multimeter and how to operate testing equipment. Oh, the most difficult thing is how to write a testing report. But fortunately, all my colleagues are very nice. They help me a lot. They teach me how to operate the machines correctly and how to write those terrible reports, which must be written in English! What about you? Miss you, Anna. Hope I can meet you very soon.

    Good day!

Love, May Lin

Answer to May Lin

Dear ____________,

Glad to ____________________________.

Very surprised that you are working at an _________________. But I know you have studied ____________________ in college. So I think that job is really suitable for you. My brother Tom now works at _______________ department, too. You two can share words when meeting next time. After you went, I found another roommate. Her name is Sue. She is pretty good. She is studying economics at university. She is now busy with her graduate_______________.

And she likes the books which you left for me. When will you come back? Call me, anytime. ____________________ to your mother and your cat. I know you miss them so much.

Nice weekend!

Love, ____________

Read the following sentences, and assort these sentences into the corresponding box.

Please give me the blueprint.

I need the manual of this CNC machine center.

I need check the processing card.

Sorry, I need your guarantee card to check your purchase date, please.

What can I do for you? What's wrong with your TV?

Let me mill the surface.

OK, I will check main circuit again.

This program is for manufacturing the curve.

Please assort the sentences above.

Miller	Customer service	Programmer	Technician

Reading Comprehension

Different trades, different jobs.

Part of a CNC programmer's job might include adjusting computer-aided drafting software simulators.

CNC programmers translate blueprint schematics into command inputs for CNC machines.

A CNC operator is someone who operates a computer numerical control machine.

A CNC operator must be able to program the machine to perform the task needed and monitor the work, making all necessary adjustments.

What is the difference between CNC operator and CNC programmer?

CNC Programmer	CNC Operator

 Tools

Look at the pictures. Please match the names in the box with the given tools.

multimeter	hammer	spanner
_____	_____	_____
power drill	screwdriver	plier
_____	_____	_____
wire stripper	vernier calliper	
_____	_____	

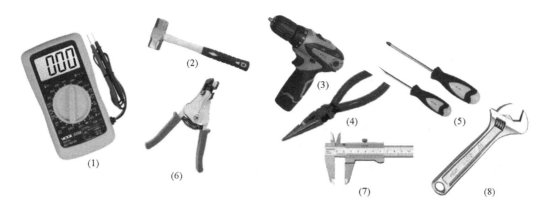

(1)　(2)　(3)　(4)　(5)　(6)　(7)　(8)

Please make sentences with the words or phrases in the table.

A multimeter	is for	cutting	wires.
A hammer		making	holes.
A spanner		tightening	nuts.
A power drill		measuring	metal.
A screwdriver		loosening	electrical circuits.
A hacksaw		stripping	work pieces.
A pair of pliers		breaking	components.
A pair of wire strippers			joined objects.
A pair of vernier callipers			

A multimeter is for _____

_____

_____

_____

_____

_____

_____

_____

Dialogue

May Lin and Li Jun are talking about a digital multimeter. Please read the dialogue.

Li Jun: May, I heard that we had bought a new digital multimeter.

May Lin: Yes. It's more convenient for measurement.

Li Jun: Can you use that one? I only have used a handy multimeter.

May Lin: Me too. But I think it isn't very difficult for learning. Let's try. Here is the manual for the multimeter. First, put on the power button.

Li Jun: Yes, I have done it.

May Lin: And then choose the measure type, current, voltage or resistance. It will display the corresponding choice on the screen.

Li Jun: Oh, I chose voltage. Then , yes , it displayed. It is not very difficult for using. It is the same as the handy one.

May Lin: Maybe. The manual said that this machine has many functions. But we haven't used it now. We need time to research how to use these functions.

Look at this multimeter. And learn the new words or phrases.

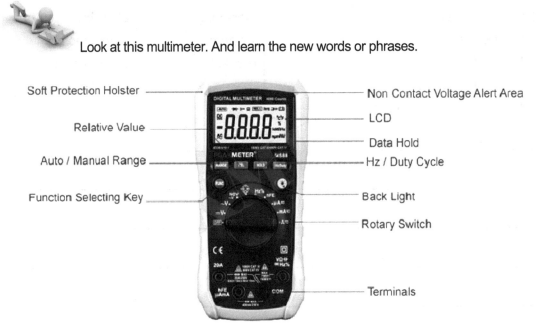

Please make a dialogue with your partner about how to use a handy multimeter to test a loop.

A: Could you tell me how to use a handy multimeter to test circuit voltage?

B: Look, a multimeter need connect a stick. Usually, we have two kinds of sticks, one is red and the other is black. You can find the sticks in the cabinet.

A: Are these sticks right side?

B: Yes, they are. And then, connect the red stick into the voltage terminal, and the black stick into the COM terminal. Like this.

A: What is the rotary switch?

B: The rotary switch is for choosing...If you want to test an AC loop's voltage, 500V/AC is the best choice.

A: But the highest voltage of my circuit is AC24. Do I need to use 500V/AC?

B: ...

A: How to read the voltage?

B: ...

A: Must I put back to the cabinet after I finished?

B: Of course. And leave your signature on this manual, which means you have used the instrument.

A: Thank you so much.

B: You are welcome. I am very glad to serve you.

 请和教师交流结果。

教师签名：＿＿＿＿＿＿＿＿

微课资源

扫一扫：获取相关微课视频。

1.1 Introduce Myself-1

1.1 Introduce Myself-2

1.1 Introduce Myself-3

1.1 Introduce Myself-4

1.1 Introduce Myself-5

1.1 Introduce Myself-6

Module 2　Electric Technology

2.1　Electric Circuit (1)

Task Content: May Lin now is working for an electric company. Her company sells electricity equipment. Now, she is learning some kinds of electricity equipment. Please learn with her.

 Please read the text, and answer the questions.

A circuit breaker is an automatically operated electrical switch designed to protect an electrical circuit from damage caused by overload or short circuit. Its basic function is to detect a fault condition and interrupt current flow. Unlike a fuse, which operates once and then must be replaced, a circuit breaker can be reset (either manually or automatically) to resume normal operation. Circuit breakers are made in varying sizes, from small devices that protect an individual household appliance up to large switch gear designed to protect high voltage circuits feeding an entire city.

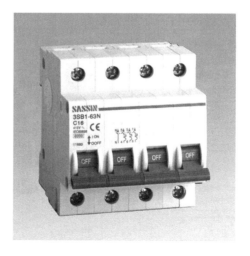

Questions

1. 这是一个什么设备？主要完成什么功能？

   _____

2. 和熔断器相比，这个装置具有什么优势？

   _____

3. 除了基本功能外，这个装置还有什么辅助功能？

   _____

This picture is the inside of the circuit breaker. Please fill in the blanks about each part's name of the circuit breaker with the given words or phrases in the box.

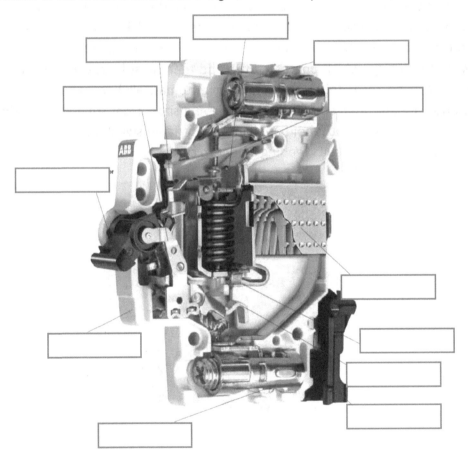

| fixed contact | moving contact | tripping lever | operator |

fixed contact　　moving contact　　tripping lever　　operator

marker　　spring　　arc chamber　　electro-magnetic protection

upper terminal　　lower terminal　　thermal protection-bimetal

Extend Knowledge

To choose an appropriate circuit breaker, you need to decide two main parameters. First, you must ensure the rate current is bigger than the total load current. To avoid ordinary use, the loop can be cut by the breaker. Second, the short circuit break current can cut the short circuit accident without taking a fire. The breaker's short circuit current must be bigger than the loop's short circuit current.

A conventional system

With three poles a circuit-breaker equipped with a magnetic only trip units for protection against short-circuits, a thermal relay for protection against overload and phase failure or imbalance, and a contactor to operate the motor.

An advanced protection system

It integrates all the protection and monitoring functions, and a contactor for operating the motor, in the circuit-breaker itself.

快速阅读：请在 3 分钟内阅读完上文，然后回答传统保护系统和现代保护系统的主要区别是什么。请不要借助任何字典或辅助工具。

思考一下，现代化的保护设备还应具备哪些特性。

_____

_____

_____

Reading Comprehension

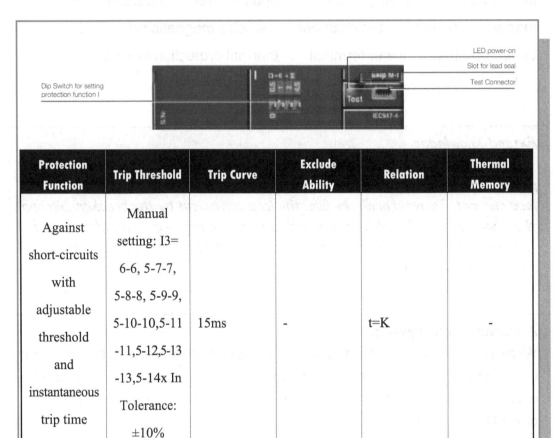

Protection Function	Trip Threshold	Trip Curve	Exclude Ability	Relation	Thermal Memory
Against short-circuits with adjustable threshold and instantaneous trip time	Manual setting: I3= 6-6, 5-7-7, 5-8-8, 5-9-9, 5-10-10,5-11 -11,5-12,5-13 -13,5-14x In Tolerance: ±10%	15ms	-	t=K	-

请阅读上述电动机保护断路器设置说明，并回答下列问题。

1. 上文主要是设置什么保护参数？

_____

2. 如果要求动作电流为 8A，应该如何操作？

_____

What Cause a Circuit Breaker to Trip on a Clothes Dryer?

The circuit breakers found in your service panel are designed to protect property and equipment against short circuits and constant current overload. Short circuits happen when a hot wire come into contact with a grounded conductor or surface. Circuit overload on a clothes dryer can be caused by a current draw that is higher than normal. A circuit breaker also can be tripped as a result of bad bearings, a defective start switch or bad motor windings.

A Bad Start Switch, Motor Winding or Shorted Heating Element

A start switch stuck in the "Run" position will cause the start winding from being connected when the motor starts, which will cause the motor to draw an exceptional high current for a prolonged period of time, causing the breaker to trip open. An open circuit in either the start or run windings will result in a high current draw and a tripped breaker. A heating element that has failed and gone to ground will also trip a circuit breaker.

Bad Bearings

Motor bearings can become worn from natural usage and can become gummed up with dirt, causing the motor shaft to bind. Badly worn bearings or severely gummed bearings will keep a motor from reaching normal operating speed. This, in turn, makes the motor draw a higher current, tripping open the circuit breaker.

Source:https://homeguides.sfgate.com/

Work with your partner.

1. Please find three main purposes for a circuit breaker to trip on the clothes dryer?

_____

2. Why does the clothes dryer switch off when it has been opened?

_____

3. What is the main purpose of the bearings?

_____

 请和教师交流结果。

教师签名：_____

2.2 Electric Circuit (2)

Task Content: May Lin and Peter Karl are talking about how to check a motor circuit. Please learn with them.

Please read the text, and answer the questions.

Manual motor starters are electromechanical devices for motor and circuit protection. These devices offer local motor disconnecting means, manual ON/OFF control, and protection against short circuits, overload, and phase loss conditions. Manual motor protection saves cost, panel space, and ensures fast and reliable short-circuit protection by reacting within milliseconds. Close coupling adapters are available for combination with ABB contactors.

The manual motor starters protect the load and the installation against short—circuits and overload. They are three-pole protection devices with thermal tripping elements for overload protection and electromagnetic tripping elements for short-circuit protection. Furthermore, they provide a disconnecting function for safe isolation of the installation and the supply and can be used for the manual switching of loads.The manual motor starters have a setting scale in amperes, which allows for direct adjustment of the device without any additional calculation. In compliance with international and national standards, the setting current is the rated current of the motor and not the tripping current (i.e. no tripping at 1.2 x I; I = setting current).

Functional Description

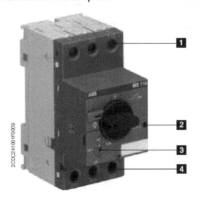

1 Terminals 1L1, 3L2, 5L3

2 Test function

3 Current setting range
Adjustable current setting for overload protection

4 Terminals 2T1, 4T2, 6T3

Source: ABB 技术资料

Questions

1. What is this?

_____

2. What is its use?

_____

3. How to choose an appropriate starter?

_____

Translation

These devices offer local motor disconnecting means, manual ON/OFF control, and protection against short circuits, overload, and phase loss conditions. Manual motor protection saves cost, panel space, and ensures fast and reliable short-circuit protection by reacting within milliseconds.

_____

_____

_____

_____

_____

Translate the new words or phrases in the word box into Chinese.

electromechanical	short circuit	overload
_____	_____	_____
phase loss	terminal	thermal tripping
_____	_____	_____

Dialogue

May Lin: Karl, I find there is something wrong with the motor.

Peter: What's the problem?

May Lin: The motor doesn't work, and has some strange sound.

Peter: Have you checked the thermal tripping?

May Lin: Yes, I have checked the thermal tripping. But the thermal tripping has no action. It must be something else.

Peter: OK. Let's check the phase loss. I'll take the multimeter.

May Lin: OK. Here is the main circuit. Phase AB is OK, no problem.

Peter: Then, the phase AC.

May Lin: The phase AC...Oh, no action. The phase AC must be lost.

Peter: Yes, we find the problem.

May Lin: But we still don't know which equipment is wrong.

Peter: Be patient. I check the wire and the contactor now. The wire is OK. Next...the contactor. Oh, look. It is the problem of the contactor. The contactor's main contact is welding.

May Lin: You are so good. I will change a new contactor. Thank you very much.

Peter: It's my pleasure.

Listening Comprehension

This is an introduction of a self-driving car. Please listen to the audio and decide whether each of following sentences is True(T) or False(F).

1. It is a real product from Google.   ☐T ☐F

2. The car has the advantage of its safety.   ☐T ☐F

3. The car has driven about a thousand miles and never have any accident.   ☐T ☐F

4. The car accident is the first reason for young's death.   ☐T ☐F

5. We waste nearly one hour each day on traffic jams that can be avoided by driverless cars.

☐T ☐F

> Do you realize that we could change the capacity of highways by a factor of two or three if we didn't rely on human precision on staying in the lane — improve body position and therefore drive a little bit closer together on a little bit narrower lanes, and do away with all traffic jams on highways? Do you realize that you, TED users, spend an average of 52 minutes per day in traffic, wasting your time on your daily commute? You could regain this time. This is four billion hours wasted in this country alone. And it's 2.4 billion gallons of gasoline wasted.

Source:www.ted.com/talks/sebastian_thrun_google_s_driverless_car

上述文字主要描述了该汽车的什么优势？

_____

_____

_____

_____

 请和教师交流结果。

教师签名：_____

2.3　Electric Circuit (3)

Task Content: May Lin got a new project for designing a new household electric system. There are some components for household electric.

Please read the text, and answer the questions.

RCDs are designed to disconnect the circuit if there is a leakage current. By detecting small leakage currents (typically 5～30 mA) and disconnecting quickly enough (<300 ms), they may prevent electrocution. They are an essential part of the automatic disconnection of supply (ADS), i.e. switching off when a fault develops, rather than relying on human intervention, one of the essential tenets of modern electrical practice. There are also RCDs with intentionally slower responses and lower sensitivities, designed to protect equipment or avoid starting electrical fires, but not to disconnect unnecessarily the equipment with greater leakage currents in normal operation. To prevent electrocution, RCDs should operate between 25 and 40 milliseconds with any leakage currents (through a person) of greater than 30 mA, before electric shock can drive the heart into ventricular fibrillation, the most common cause of death through electric shock. By contrast, conventional circuit breakers or fuses only break the circuit when the total current is excessive (which may be thousands of times as big as the leakage current that an RCD responds to). A small leakage current, such as through a person, can be a very serious fault, but would probably not increase the total current enough for a fuse or circuit breaker to break the circuit, and certainly not do so fast enough to save a life.

Type Designation

GS□□ S □ □ □

———— Rate residual current (mA)

———— Rated current (A)

———— Tripping Characteristics: C,D

———— Pole: NA

———— Type: Electronics Type

Questions

1. What is the main purpose of an RCD?

_____

2. How much leakage current can cause an RCD to operate?

_____

3. How much leakage current will cause a serious accident for a man?

_____

Translation

A small leakage current, such as through a person, can be a very serious fault, but would probably not increase the total current enough for a fuse or circuit breaker to break the circuit, and certainly not do so fast enough to save a life.

_____

_____

_____

_____

_____

Translate the new words or phrases in the word box into Chinese.

leakage current	millisecond	electrocution
_____	_____	_____
electric shock	by contrast	ventricular fibrillation
_____	_____	_____

Do you know these equipment?

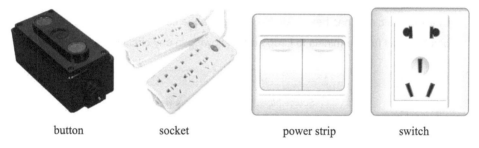

| button | socket | power strip | switch |

 Extend Reading

These guidelines detail the necessary precautions that should be taken by all personnel to minimize their exposure to electrical sources of energy.

- Electrical repair work or diagnostic work on electrical equipment shall only be performed by personnel that are qualified to perform this task. Qualified personnel must be documented by the Safety Department.
- Instruction manuals from manufacturers detail what work can be performed by specific individuals. Consult the manuals for instruction. If in doubt, seek additional help.
- Portable electrical hand tools must be grounded or double insulated.
- Temporary portable lighting used in damp and/or hazardous locations and confined areas with low ground resistance must be operated at a maximum of 12 volts.

- All electrical cords and cables are covered or elevated to protect them from damage and to eliminate tripping hazards.
- Qualified electricians are the only employees authorized to repair electrical equipment. Field repairs or tampering with any electrical equipment by unauthorized persons is not tolerated.
- No work should be performed hot regardless of voltage. When it becomes necessary to work on energized lines or equipment, the task is reviewed and approved by the company's electrical supervisor. When doing authorized hot jobs, approved rubber electrical gloves, blankets, mats and other protective equipment must be used.
- Distribution panels must be dead front type, covering hot terminals and properly constructed and grounded. High voltage (600 volts or more) must be properly protected and identified by approved signs.
- Ground fault circuit protection must be used on all electrical systems, 220 volts receptacle outlets, extension cords and equipment connected by electrical cords and plugs. Distribution panels supplied at the site incorporate RCD device for 220 volt usage.
- Be certain the circuit has been switched off before the connection is broken.
- Do not bypass fuse terminals to keep current flowing in any circuit.

Source:https://wenku.baidu.com/view/bccf9a34f524ccbff021846f.html?from=search

Please translate the following sentences into Chinese. The vocabulary box will help you.

guidelines	指导意见	diagnostic work	诊断工作
electrical equipment	电气设备	consult	请教
tamper	篡改	energize	给定电压
electrical supervisor	电气专家	rubber electrical gloves	绝缘手套
blanket	绝缘毯	mat	绝缘垫
ground fault circuit protection	接地保护		
distribution panel	接线板		
terminal	接线端子		
bypass	跨线、短接		

1. Temporary portable lighting used in damp and/or hazardous locations and confined areas with low ground resistance must be operated at a maximum of 12 volts.

_____

_____

2. Distribution panels must be dead front type, covering hot terminals and properly constructed and grounded. High voltage (600 volts or more) must be properly protected and identified by approved signs.

_____

_____

_____

3. Be certain the circuit has been switched off before the connection is broken. Do not bypass fuse terminals to keep current flowing in any circuit.

_____

_____

请和教师交流结果。

教师签名：_____

2.4 Basic Terminology

Task Content: As a student of engineering, there are some Basic Terminologies what you should be familiar with. Let's learn them together.

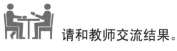

Please fill in the form below.

前缀	符号	数值	中文名称	例　词
giga–	G			Gigahertz(GHz)
meg(a)–	M			megameter, megahertz
kilo–	k			kilometer(km), kilovolt(kV), kilowatt(kW)
hecto–	h			hectogram
deca–	da			decameter

续表

前缀	符号	数值	中文名称	例　词
deci-	d			decimeter
centi-	c			centimeter
milli-	m			milliampere(mA), milliammeter
micro-	μ			microfarad(μF)
nano-	n			nanometer(nm)
pico-	p			picocoulomb(pC), picofarad(pF)

Questions

1. Do you know them?

Ohm ＿＿＿＿＿＿　　　Ampere ＿＿＿＿＿＿

Volt ＿＿＿＿＿＿　　　Hertz ＿＿＿＿＿＿

Watt ＿＿＿＿＿＿　　　Kelvin ＿＿＿＿＿＿

Coulomb ＿＿＿＿＿＿

对一些科学家、发明家所发现或发明的理论、方法、定理或定律、东西和物品等，用它们的名字组成新词。

2. Translate the following acronyms into Chinese.

AC　　—Alternating Current

DC　　—Direct Current

CD　　—Compact Disc

LED　 —Light Emitting Diode

ID　　—Identification Card

IP　　—Internet Protocol

UPS　 —Uninterruptible Power Supply

HV　　—High Voltage

HVDC —High Voltage Direct Current

IEE　 —Institution of Electrical Engineers

IEEE　—Institute of Electrical and Electronics Engineers

3. Do you know other acronyms in electrical engineering? Please write them down.

_____

_____

_____

During studying, the mathematics is necessary for every student. Do you know how to read the symbols and formulas in English?

x^2	:	x square, x squared, the square of x, the second power of x, x to the second power
y^3	:	y cube, y cubed, the cube of y, the third power of y, y to the third power
a^n	:	the nth power of a, a to the n power
$\sqrt{3}$	:	the square root of three
$\sqrt[3]{a}$	:	the cube(third) root of a
$\sqrt[n]{a}$	:	the nth root of a
0.1	:	zero point one, point one, o point one, one tenth, decimal one
0.01	:	point zero one
10.35	:	ten point three five
1/2	:	(one)half
1/3	:	a third
1/4	:	one quarter
1/8	:	a eighth
2/3	:	two-thirds, two over three, two divided by three
3/4	:	three-fourths, three quarters
3%	:	three percent
0.2%	:	point two percent
5‰	:	five per mill

Let's do some exercises.

$\sqrt[3]{9}$

0.25‰

4/5

20%

9.24

4^2

Work with your partner to fill in the form.

A+B	
A-B	
A×B	
A÷B	
A±B	
A=B	
A≠B	
A≈B	
A>B	
A<B	
A≥B	
A≤B	

 请和教师交流结果。

教师签名：＿＿＿＿＿＿

2.5 Electric Motor

Task Content: Peter Karl now is choosing the electric motor. And a few old motors need repairing. Please learn with him.

Please learn the data plate of electric motors and answer the questions.

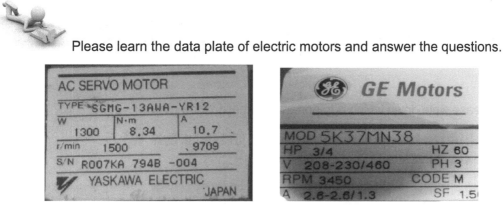

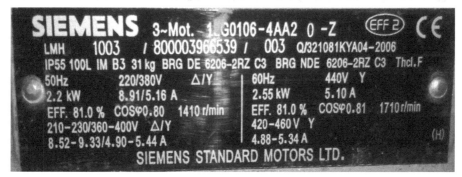

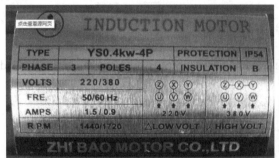

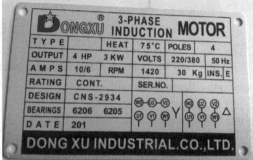

 Please write down which is important information for a motor.

(1) 上图中哪个铭牌不是电动机？推测一下可能是什么设备。

(2) 请判断上述设备中的各电动机的型号。

(3) 请判断上述设备中各电动机功率。

(4) 判别上述电动机的类型，如果是三项交流电机，则读出它的额定工作频率。

(5) 请判断上述电动机的额定转速和额定电压。

Reading Comprehension

YE2 series high-quality three-phase ac electric motors

A YE2 series high-quality three-phase ac electric motor is a kind of totally enclosed fan-cooled squirrel-cage induction motors, it was developed in 1990' in China, the design of YE2 series ac electric motors absorbed advanced technology of electric motors around the world. YE2 series ac electric motors have lots of features such as high efficiency and energy saving, large starting torque, excellent performance, low noise and vibration, compact structure, high reliability, easy operation, etc.

Dimensions and rated output of YE2 series ac electric motors are in conformity with international standard IEC 60072-1:1991, so do Y series induction motors and YE3 series acelectric motors. While an efficiency class of YE2 series ac induction motors is in accordance with Chinese National standard GB 18613-2006 class III, on request of the user, we supply YE2 series ac motors in line with IEC 60034-30 High Efficiency standard (IE2) as well.

YE2 series three-phase ac electric motors are widely used as drive mechanism in various kinds of machineries like machining tools, pumps, fans, compressors, gear boxes, etc., and can also be used in transportation, agricultural machinery, printing industry and food industry.

1. What is this?

_____

2. List the advantage of the motor.

_____

3. Which workplace is this motor used in?

_____

Translation

A YE2 series high-quality three-phase ac electric motor is a kind of totally enclosed fan-cooled squirrel-cage induction motors, it was developed in 1990 in China, the design of YE2 series ac motors absorbed advanced technology of electric motors around the world. YE2 series ac electric motors have lots of features such as high efficiency and energy saving, large starting torque, excellent performance, low noise and vibration, compact structure, high reliability, easy operation, etc.

_____

_____

_____

_____

_____

_____

_____

Translate the new words or phrases in the word box into Chinese.

torque	squirrel-cage induction motor
_____	_____
pump	fan agricultural machinery
_____	_____ _____

Dialogue

May Lin: Hi, Peter. Tell me something about the DC motor, please. I have learned much about the AC Motor in my college. But I know little about the DC motor.

Peter: No problem. The DC motor is more and more popular nowadays.

May Lin: What is the difference of these two kinds of machine?

Peter: The difference is clear. The DC motor uses direct current. The AC motor uses alternating current.

May Lin: Yes, I know that. I mean their different advantages and disadvantages.

Peter: Ok. The DC motor in past usually needs brush to exchange the current direction in the rotor. When the current goes higher, it will have the spark. The brush is not safe. But the DC motor has a wider speed range than the AC motor. Nowadays, we invented brushless DC motors to conquer the disadvantage of the spark in exchange. So the DC motor becomes more and more popular.

May Lin: Brushless DC motors? Do they need additional devices for exchanging the current directions?

Peter: Definitely. Usually, we called that "driver program". And every brush of DC motors needs a driver to ensure its running.

May Lin: It will be very difficult to program a brushless motor?

Peter: No, you don't worry about this. When you buy a brushless motor, you will get this program. It will be sold with the DC motor.

May Lin: That is great! So convenient. And what about the AC motor?

Peter: Most AC motors are used in fans. They are cheaper and needn't additional program for operating. The AC motor has a huge advantage for its small torque. In heavy-duty machines usually you will not see AC motors.

May Lin: Thank you. Now I know the DC motor. It has more advancement. Sorry for
disturbing you so long.

Peter: No problem. Now, let's go for coffee.

May Lin: Certainly, it's coffee time. Today, my treat.

Peter: Thank you. You're so kind.

Please read the dialogue and decide whether each of the following statements is True(T)
or False(F).

1. The DC motor uses alternating current.   □T □F
2. The AC motor needn't additional driver programs.   □T □F
3. The DC motor becomes more and more popular.   □T □F
4. The AC motor is used in heavy-duty machines.   □T □F
5. The DC motor has the disadvantage of the brush. Maybe it will cause a fire. □T □F

Reading Comprehension

Look at the following paragraph. It is a kind of household application. Please guess what it is?

- Room Size: 580 square feet
- Clean Air Delivery Rate: 375 cubic feet per minute
- Air changes per hour: 5
- Airflow: 90~415 cubic feet per minute
- Height × Width × Depth: 26 × 20 × 13 inches
- Product weight: 35 pounds
- Energy consumption: 35~120 watts
- Noise level: 32~66 decibels
- Electronic sensors, remote, on/off timer: No
- Filter replacement indicator: Yes
- Speed control options: 1 - 2 - 3

Writing

Please read the information of a smart blub, which can be used in an ordinary family. And write an introduction of that smart bulb.

Play with light and choose from 16 million colors to instantly change the look and atmosphere of your room. Set the scene effortlessly with one touch of a button. Use a favourite photo and relive that special moment with splashes of light. Save your favourite light settings and recall them whenever you want with the tap of a finger.

With iOS and Android Apps you can control your lights remotely wherever you are. Check if you have forgotten to switch your lights off before you leave your home, and switch them on if you are working late.

 请和教师交流结果。

教师签名：_____

2.6　Electronics Circuit

Task Content: Li Jun now is checking the electronics components. Please help him.

Please read the Manual and answer the following questions.

May 1998

National *Semiconductor*

LM337L
3-Terminal Adjustable Regulator

General Description

The LM337L is an adjustable 3-terminal negative voltage regulator capable of supplying 100 mA over a 1.2V to 37V output range. It is exceptionally easy to use and requires only two external resistors to set the output voltage. Furthermore, both line and load regulation are better than standard fixed regulators. Also, the LM337L is packaged in a standard TO-92 transistor package which is easy to use.

In addition to higher performance than fixed regulators, the LM337L offers full overload protection. Included on the chip are current limit, thermal overload protection and safe area protection. All overload protection circuitry remains fully functional even if the adjustment terminal is disconnected.

Normally, only a single 1 μF solid tantalum output capacitor is needed unless the device is situated more than 6 inches from the input filter capacitors, in which case an input bypass is needed. A larger output capacitor can be added to improve transient response. The adjustment terminal can be bypassed to achieve very high ripple rejection ratios which are difficult to achieve with standard 3-terminal regulators.

Besides replacing fixed regulators, the LM337L is useful in a wide variety of other applications. Since the regulator is "floating" and sees only the input-to-output differential voltage, supplies of several hundred volts can be regulated as long as the maximum input-to-output differential is not exceeded.

Also, it makes an especially simple adjustable switching regulator, a programmable output regulator, or by connecting a fixed resistor between the adjustment and output, the LM337L can be used as a precision current regulator. Supplies with electronic shutdown can be achieved by clamping the adjustment terminal to ground which programs the output to 1.2V where most loads draw little current.

The LM337L is available in a standard TO-92 transistor package and a SO-8 surface mount package. The LM337L is rated for operation over a −25°C to +125°C range.

For applications requiring greater output current in excess of 0.5A and 1.5A, see LM137 series data sheets. For the positive complement, see series LM117 and LM317L data sheets.

Features

- Adjustable output down to 1.2V
- Guaranteed 100 mA output current
- Line regulation typically 0.01%/V
- Load regulation typically 0.1%
- Current limit constant with temperature
- Eliminates the need to stock many voltages
- Standard 3-lead transistor package
- 80 dB ripple rejection
- Output is short circuit protected

Connection Diagram

Bottom View

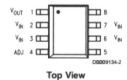

Top View

Order Number LM337LM or LM337LZ
See NS Package Number M08A or Z03A

Please answer the following questions.

1. What is the use of LM337L?

2. What is the output voltage range of LM337L?

_____

3. How much is output current of LM337L?

_____

4. What is the additional protection of LM337L?

_____

Please draw and translate the connection diagram.

Regulator with Trimmable Output Voltage

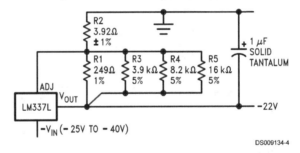

DS009134-4

Trim Procedure:

— If VOUT is −23.08V or bigger, cut out R3 (if smaller, don't cut it out).

— Then if VOUT is −22.47V or bigger, cut out R4 (if smaller, don't).

— Then if VOUT is −22.16V or bigger, cut out R5 (if smaller, don't).

This will trim the output well within 1% of −22.00 VDC, without any of the expense or trouble of a trim pot (see LB-46). Of course, this technique can be used at any output voltage level.

1. 上图电路的主要功能是什么？

_____

2. R3、R4、R5 主要起到什么作用？

_____

3. 在所有电阻都不切断的状态下，这个电路输出是多少？

_____

4. 此电路采用什么滤波方式？

_____

Translate the new words or phrases in the word box into Chinese.

operational	amplifier	input and output
_____	_____	_____
resistance	inductance	capacitance
_____	_____	_____
filter	regulator	transistor
_____	_____	_____

 Extend Reading

Operational Amplifiers

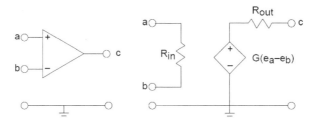

The op-amp has four terminals to which connections can be made. Inputs attach to nodes **a** and **b**, and the output is node **c**. As the circuit model on the right shows, the op-amp serves as an amplifier for the difference of the input node voltages.

Op-amps not only have the circuit model , but also their element values are very special.

- The **input resistance**, R_{in}, is typically **large**, on the order of 1MΩ.
- The **output resistance**, R_{out}, is **small**, usually less than 100Ω..
- The **voltage gain**, G, is **large**, exceeding 10^5.

1. 请思考，上文中谈到的理想运放具有什么特性？请和同组成员讨论，并且不要使用字典。

_____

2. 请不借助字典理解上文，写出理想运放的输入、输出电路和理想放大倍数。

_____

Group Task

Use the material of Thevenin theorem to solve the problems blow. Make a presentation in your classroom.

> **Thevenin Theorem.**
> Any linear electrical network with voltage and current sources and resistances only can be replaced at terminals A-B by an equivalent voltage source Vth in series connection with an equivalent resistance Rth.
> **Maximum Power Transfer Theorem**
> It obtains maximum external power from a source with a finite internal resistance, the resistance of the load must equal the resistance of the source.

Problem: Big is Beautiful
Sammy wants to choose speakers that produce very loud music. He has an amplifier and notices that the speaker terminals are labeled "8Ω source."
(a) What does this mean in terms of the amplifier's equivalent circuit?
(b) Any speaker Sammy attaches to the terminals can be well-modeled as a resistor. Choosing a speaker amounts to choosing the values for the resistor. What choice would maximize the voltage across the speakers?
(c) Sammy decides that maximizing the power delivered to the speaker might be a better choice. What values for the speaker resistor should be chosen to maximize the power delivered to the speaker?

 请和教师交流结果。

教师签名：_____

2.7　Electric Element (1)

Task Content: Peter is a new student of Electrical Engineering. He needs to learn some basic electrical elements, such as what they are, what they can do in the circuit, how many types there are...

Please read the text, and answer the questions.

Resistance may be thought of as an opposition to current flow; the higher the resistance is, the lower the current will flow.

The resistance of an electrical conductor depends on 4 factors, which are as follows.

(a) The first is the length of the conductor. Resistance (R) is directly proportional to length (l) of a conductor, i.e. $R \propto l$.

(b) The second is the cross-sectional area of the conductor. Resistance (R) is inversely proportional to a cross-sectional area (a) of a conductor, i.e. $R \propto 1/a$.

(c) The third is the type of materials.

(d) The fourth is the temperature of the materials.

Questions

1. What is this?

_____

2. What is the function of this element?

_____

3. Write down four factors of the element.

_____

_____

机电专业英语项目化教程(微课版)

_____

_____

This picture shows the color-coding scheme of four-band resistors.

Color	1st band	2nd band	3rd band (multiplier)	4th band (tolerance)	Temp. Coefficient
Black	0	0	$\times 10^{0}$		
Brown	1	1	$\times 10^{1}$	±1% (F)	100 ppm
Red	2	2	$\times 10^{2}$	±2% (G)	50 ppm
Orange	3	3	$\times 10^{3}$		15 ppm
Yellow	4	4	$\times 10^{4}$		25 ppm
Green	5	5	$\times 10^{5}$	±0.5% (D)	
Blue	6	6	$\times 10^{6}$	±0.25% (C)	
Violet	7	7	$\times 10^{7}$	±0.1% (B)	
Gray	8	8	$\times 10^{8}$	±0.05% (A)	
White	9	9	$\times 10^{9}$		
Gold			$\times 10^{-1}$	±5% (J)	
Silver			$\times 10^{-2}$	±10% (K)	
None				±20% (M)	

Please write down the value of each resistor below.

Translate the new words or phrases in the word box into Chinese.

tolerance accuracy	significant	multiplier
_____	_____	_____
thermal coefficient	color-coding scheme	
_____	_____	

Reading Comprehension

Four-band identification is the most commonly used color-coding scheme on resistors. It consists of four colored bands that are painted around the body of the resistor. The first two bands encode the first two significant digits of the resistance value, the third is a power-of-ten multiplier or number-of-zeroes, and the fourth is the tolerance accuracy, or acceptable errors, of the value. The first three bands are equally spaced along the resistor; the spacing to the fourth band is wider. Sometimes a fifth band identifies the thermal coefficient, but this must be distinguished from the true five-color system, with three significant digits.

Work with your partner.

1. What is the four-band identification of the resistor?

2. Make a dialogue for how to identify different resistors.

 请和教师交流结果。

教师签名：＿＿＿＿＿＿＿＿

2.8 Electric Element (2)

Task Content: Peter is a new student of Electrical Engineering. He needs to learn some basic electrical elements, such as what they are, what they can do in the circuit, how many types they are...

Please read the text, and answer the questions.

Electrical energy can be stored in an electric field. The device capable of doing this is called a capacitor or a condenser. Capacitors consist of two metal plates that are separated by all insulating materials. If a battery is connected to both plates, an electric charge will flow for a short time and accumulate on each plate. If the battery is disconnected，the capacitor retains the charge and the voltage associated with it.

The ability of a capacitor to store electrical energy is termed capacitance. The capacitance is directly proportional to the dielectric constant of the materials and to the area of the plates and inversely to the distance of the plates. It is measured in farads .When a changer of one volt per second across it causes the current of one ampere to flow, the condenser is said to have the capacitance of one farad. However, farad is too large a unit to be used in radio calculation, so microfarad (10^{-6} farads) and the picofarad (10^{-12} farads) are generally used.

The main types of the capacitor include variable air, mica, paper, ceramic, plastic, titanium oxide and electrolytic.

Questions

1. What is called a condenser? Which elements do condensers consist of?

_____

_____

_____

2. How to define one farad of the capacitance?

_____

3. What is the function of the capacitor?

_____

4. Write down main types of the capacitor.

_____

_____

Please read the text, and answer the questions.

Inductors consist of a conducting wire wound into the form of a coil. When a current passes through the coil, a magnetic field is set up around it that tends to oppose rapid changes in current intensity. All coils have inductance.

A coil of many turns will have more inductance than one of few turns. Also if a coil is placed on an iron core, its inductance will be greater than it was without the magnetic core. The unit of inductance is the henry. A coil has inductance of one henry if an induced emf of one volt is induced in the coil when the current through it changes at the rate of one ampere per second. Values of inductance used in radio equipment vary over a wide range.

Questions

1. What are inductors made up of?

_____

_____

2. How to define one henry of inductance?

_____

3. What is the difference between the inductor with the magnetic core and that without the magnetic core?

_____

Let's do some exercises.

I. Fill in the missing words according to the text.

1. This current depends on the voltage of the battery, on the_____of the sample, and on the _____of the material itself.

2. A_____is a device designed to have capacitance.

3. Ohm is used as a unit of_____.

4. The larger the emf is, the more_____the capacitor stores.

5. When a changer of one volt per second across it causes the current of one ampere to flow, the _____is said to have the capacitance of one_____.

6. The_____are made from carbon mixtures，metal films，or resistance wire and have two _____wires attached.

II. Translate the following sentences into English.

1. 可变电阻器常用来控制收音机和电视机的音量。

_____

_____

2. 电容器存储电能的能力叫电容。

_____

_____

3. 任何电感都是由线圈组成，线圈的匝数越多，电感越大。

_____

_____

4. 电容器的电容量与介质的介电常数及平板的面积成正比，与平板间的距离成反比。

_____

_____

请和教师交流结果。

教师签名：_____

2.9   Electric Element (3)

Task Content: Peter is a new student of Electrical Engineering. He needs to learn some basic electrical elements, such as what they are, what they can do in the circuit, how many types there are...

The diode is the most common element in a circuit. Do you know the following terminologies about the diode?

forward-bias	voltage drop	leakage current
_____	_____	_____
peak inverse voltage	threshold voltage	breakdown voltage
_____	_____	_____

Please read the text, and answer the questions.

A diode is an electronic component with two electrodes (connectors). It allows electricity to go through it only in one direction. It can be influenced by temperature. When the temperature increases, the cut-in voltage goes down. This makes it easier for electricity to pass through the diode.

Diodes can be used to convert alternating current to direct current (Diode bridge). They are often used in power supplies and sometimes to decode amplitude modulation radio signals (like in a crystal radio). a light-emitting diode (LED) is a type of diodes that produce light. A zener diode is like a normal diode, but instead of being destroyed by a big reverse voltage, it lets electricity through. The voltage needed for this is called the breakdown voltage or Zener voltage. Because it is built with a known breakdown voltage, it can be used to supply a known voltage.

Today, the most common diodes are made from semiconductor materials such as silicon or sometimes germanium.

Questions

1. What is the common function of a diode?

_____

_____

2. How many types of diodes are there in the text?

_____

_____

3. What is the influence of temperature?

_____

_____

4. Write down the other types of the diodes you know besides those referred to in the text.

_____

_____

Anode ▷ Cathode	Anode ▶ Cathode	Anode ▷ Cathode	Anode ▷ Cathode
Diode	Zener Diode	Schottky diode	Tunnel diode
Anode ▷ Cathode	Anode ▷ Cathode	Anode ▷ Cathode	Anode ▷ Cathode / Gate
Light-emitting diode	Photodiode	Varicap	Silicon controlled rectifier

The picture below is the data sheet of 1N4148. Please write down the features and applications of this diode.

High-speed diodes   1N4148

FEATURES

- Hermetically sealed leaded glass SOD27 (DO-35) package
- High switching speed: max. 4 ns
- General application
- Continuous reverse voltage: max. 100 V
- Repetitive peak reverse voltage: max. 100 V
- Repetitive peak forward current: max. 450 mA.

APPLICATIONS

- High-speed switching.

DESCRIPTION

The 1N4148 and 1N4448 are high-speed switching diodes fabricated in planar technology, and encapsulated in hermetically sealed leaded glass SOD27 (DO-35) packages.

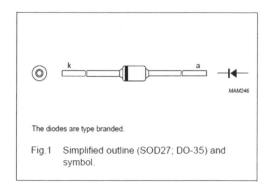

The diodes are type branded.

Fig.1   Simplified outline (SOD27; DO-35) and symbol.

MARKING

TYPE NUMBER	MARKING CODE
1N4148	1N4148PH or 4148PH
1N4448	1N4448

Features

_____

_____

_____

_____

_____

_____

_____

Applications

_____

_____

_____

There are some limiting values about 1N4148. Read the data sheet and answer the questions.

High-speed diodes   1N4148

LIMITING VALUES

In accordance with the Absolute Maximum Rating System (IEC 60134).

SYMBOL	PARAMETER	CONDITIONS	MIN.	MAX.	UNIT
V_{RRM}	repetitive peak reverse voltage		–	100	V
V_R	continuous reverse voltage		–	100	V
I_F	continuous forward current	see Fig.2; note 1	–	200	mA
I_{FRM}	repetitive peak forward current		–	450	mA
I_{FSM}	non-repetitive peak forward current	square wave; T_j = 25 °C prior to surge; see Fig.4			
		$t = 1 \mu s$	–	4	A
		$t = 1 ms$	–	1	A
		$t = 1 s$	–	0.5	A
P_{tot}	total power dissipation	T_{amb} = 25 °C; note 1	–	500	mW
T_{stg}	storage temperature		–65	+200	°C
T_j	junction temperature		–	200	°C

Note

1. Device mounted on an FR4 printed-circuit board; lead length 10 mm.

Questions

1. Can 1N4148 be used in AC circuit?

_____

2. If the circuit current is 1 A, how long will the diode be destroyed?

_____

3. What is the power dissipation of 1N4148?

_____

 请和教师交流结果。

教师签名： _____

2.10　Power Electronic Technology

Task Content: Do you know the typical power electronic component? In this lesson, we will learn the common power electronic component such as ASR, MOSFET, IGBT.

This is a parameter table of a kind of IGBT. Please read and fill in the blanks. Try your best and don't use your dictionary.

■Discontinue : IGBT Module										
Connection	Type name	Absolute Max. Ratings		Electrical Characteristics				outline	RoHS Status	Status
		VCES (V)	Ic (A)	VCE(sat) (V)Typ.	ton (us)Max.	toff (us)Max.	tf (us)Max.			
	MBN600GR12	1,200	600	2.2	1.0	1.2	0.4	N-6	N	D
	MBN400GR12	1,200	400	2.2	0.9	1.1	0.35	N-5	N	D

MBN600 额定反向最大电压 ＿＿＿＿＿＿＿＿

MBN600 额定通态平均电流 ＿＿＿＿＿＿＿＿

导通时间 ＿＿＿＿＿＿＿＿＿＿

关断时间 ＿＿＿＿＿＿＿＿＿

最小到正向导通电压 ＿＿＿＿＿＿＿＿＿＿

Read the text and decide whether each of the following statements is True(T) or False(f).

> The introduction of Power MOSFET was originally regarded as a major threat to the power bipolar transistor. This is especially true in high frequency circuits where the power MOSFET is particularly valuable due to its inherently high switching speed. On the other hand, MOSFETs have a higher state resistance per unit area and consequently a higher state loss. This is particularly true for higher voltage devices (greater than about 500 volts) which restricted the use of MOSFETs to low voltage high frequency circuits.The IGT device has undergone many improvement cycles to result in the modern Insulated Gate Bipolar Transistor (IGBT). These devices have near ideal characteristics for high voltage (> 100V) medium frequency (< 20 kHZ) applications.

1. Power MOSFET is actually a kind of transistors.　　　　　□T　　□F

2. MOSFET is due to switch the circuit with high frequency speed.　□T　　□F

3. MOSFET component's power consumption is low.   □T   □F

4. IGBT devices have ideal characteristics for high voltage and frequency.   □T   □F

5. Insulated Gate Bipolar Transistor (IGBT) is more advanced than IGT.   □T   □F

 请同学们不要使用字典，看看能不能填写上述信息。

Reading Comprehension

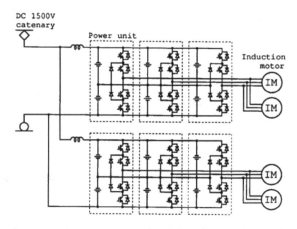

A THREE-LEVEL TRACTION INVERTER WITH IGBTs FOR EMU

Kiyoshi Nakata, Kiyoshi Nakamura, Satoru Ito
Hitachi Research Laboratory, Hitachi, Ltd.
Katsuta City, Ibaraki Prefecture, JAPAN

and Keiji Jinbo
Mito Works, Hitachi, Ltd.

TABLE I
MAIN CIRCUIT SPECIFICATIONS

Electric system	DC1500V (Maximum 1800V)
Inverter type	Three-level inverter (Neutral-point-clamped inverter)
Power device	2000V/325A IGBT module
Cooling method	Water cooling
Inverter output	800kVA x 2
Motor output	200kW x 4
Output voltage	0-1170V (at catenary voltage DC1500V)
Output frequency	0-200Hz

1. 上述设备主要完成什么功能？

_____

_____

2. 请尝试阅读参数表，并将下表翻译完整。

项目	内容
冷却方式	
输出电压	
输出频率	
电动机输出	
电气系统	
逆变类型	
电源模块	

Translate the new words or phrases in the word box into Chinese.

inverter	power bipolar transistor	high switching speed
_____	_____	_____

power electronic technology	main circuit	ideal characteristic
_____	_____	_____

rating current	transistor	solar panel inverter
_____	_____	_____

Talking with Your Partner

There is a diagram about globe solar PV inverter market. Please read and talk with your partner about the following questions. And describe the following diagram shortly.

1. The growth of this market is sharp, why?

2. Do you know which country now uses solar power?

3. Do you think we will all use solar power in the soon future?

4. Do you know the difficult points in solar power system?

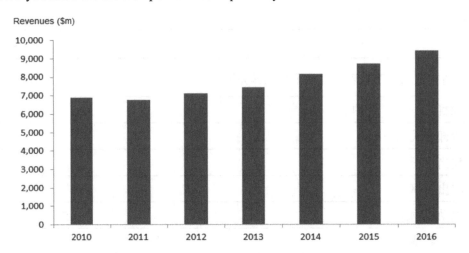

Global PV Inverter Revenues (Source: IMS Research/IHS)

 **Extend Reading**

Grid-tie inverters convert DC electrical power into AC power suitable for injecting into the electric utility company grid. The grid-tie inverter (GTI) must match the phase of the grid and maintain the output voltage slightly higher than the grid voltage at any instant. A high-quality modern grid-tie inverter has a fixed unity power factor, which means its output voltage and current is perfectly lined up, and its phase angle is within one degree of the AC power grid. The inverter has an on-board computer which senses the current AC grid waveform, and outputs voltage to correspond with the grid. However, supplying reactive power to the grid might be necessary to keep the voltage in the local grid inside allowed limitations. Otherwise, in a grid segment with considerable power from renewable sources, voltage levels might rise too much at times of high production, i.e. around noon with solar panels.

Grid-tie inverters are also designed to quickly disconnect from the grid if the utility grid goes down. This is an NEC requirement ensured in the event of a blackout, the grid-tie inverter will shut down to prevent the energy that transfers from harming any line workers who are sent to fix the power grid.

Properly configured, a grid tie inverter enables a home owner to use an alternative power generation system like solar or wind power without extensive rewiring and without batteries. If the alternative power being produced is insufficient, the deficit will be sourced from the electricity grid.

Source: Wikipedia.org

1. 电网逆变器的主要作用是什么？

2. 选用电网逆变器时应该如何匹配？

3. 电网逆变器还可以检测哪些电网问题？

 请和教师交流结果。

教师签名：＿＿＿＿＿＿＿＿＿

2.11　PLC Manual

Task Content: May Lin is learning how to connect PLC. Please read the manual and help her.

Please read the PLC Manual contents. And fill in the page number according to the following questions.

——— CONTENTS ———

Peter 想知道怎么用面板操作 PLC，那么他需要翻到

May Lin 想知道如何设置模拟量输入和数字量输入，她要去

李军想知道 PLC 电脑套件系统安装信息，他要去

他们所有的人都要学习如何设置传输参数，那么他们要找

编程组成员在写程序时要了解数据寄存器的相关信息，要去找

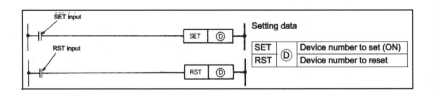

◆Function

SET

- When SET input is turned ON, the specified device turns ON.
- A device that was turned ON remains ON even if the SET input is turned OFF. It can be turned OFF with the RST instruction.

- When the SET input is OFF, the device state does not change.

RST

- When the RST input is turned ON, the specified device is as follows.

Device	Status
Y, M	The coil and contact are turned OFF.
T, C	The present value is set to 0, and the coil and contact are turned OFF.
D	The content is set to 0.

- When the RST input is OFF, the device state does not change.
- The RST (D) function is identical to the following ladder.

1. 该指令可以驱使多少种设备?

2. 该指令是如何运行的?

3. 请阅读手册的最后一段，解释如何激活数据寄存器和数据寄存器清零。

3000	PARAMETER ERROR	The content of parameter indicated by th error individual information (SD16) is incorrect.	• Read the error individual information with a programming tool, check the parameter items the correspond to the values (parameter numbers) and correct them. • Re-write the corrected parameters, reset the power or reset the inverter. • If the same error is displayed again, a CPU hardware error has occurred. Please contact your sales representative and explain the failure symptom.

错误代码 3000 表示什么？如何才能清除这个故障？

_____

_____

_____

_____

Translate the new words or phrases in the word box into Chinese!

compose	origin	automatic control system		
_____	_____	_____		
installation	car model	cycle	modify	reliability
_____	_____	_____	_____	_____
input	output	electromagnetic valve		contactor
_____	_____	_____		_____

Reading Comprehension

May Lin is now reading the history and development of PLC. Please read the text and tell her the advantage and development of PLC.

The origin of PLC is in the 1960s, auto production lines of the automatic control system are largely composed of by relay control devices.

Car retrofit directly leads to relay control devices to design and installation. With the development of production, car models update cycle is increasingly short, so relay control devices will often need to design and installation, time-consuming, fee expected, even hindered update cycle shortens. In order to change this situation, GM in 1969 with new public bidding, demanding control devices instead of relay control devices, puts forward ten tender indices, which are as follows.

1. Convenient, on-site programming can modify the program.
2. Maintenance convenience, adopt modular structure.
3. Reliability is higher than relay control device.
4. Less than relay control devices.
5. The data can be directly into management computer.
6. Costs can be competitive with relay control devices.
7. Input is communication 115V.
8. The output of ac 115V, 2A above, can directly drive electromagnetic valve, contactor, etc.
9. As to its expanding, the original system has small change.
10. The user program memory capacity at least could be extended to 4K.

Why do we use PLC?What about the future of PLC? Please match the sentences in the following two boxes.

(1)PLC has been tasked with is...

(2)Latest technology gives PLC a faster, ...

(3)These advances have also allowed PLC to expand its portfolio and take on new tasks like...

(4) Ability to place PLC in closer proximity to real world devices and communicate back to...

(5)Even simple motion control previously required...

(6)They have to interface with bar code scanners and printers, as well as temperature....

(A)... to more sophisticated control platforms in order to meet system requirements.

(B)... more powerful processor with more memory at less cost.

(C) ... motion control.

(D)... and analog sensors. They need multiple protocol support to be able to connect with other devices in the process.

(E)....communications,data manipulation and high-speed motions without giving up the rugged and reliable performance expected from industrial control equipment.

(F)... other system controls in a main panel.

 请和教师交流结果。

教师签名： ＿＿＿＿＿＿＿＿＿＿＿＿

2.12 Robot and Industry 4.0

Task Content: May Lin wants to learn robot systems applied in industry manufacture. Please learn with her.

New Robot, New Technology

Look at the pictures. And learn the new words or phrases in the box.

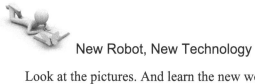

ABB Yumi @ 2015

- Dual arm with 7-DOF
- Human-robot collaborative
- Zero-foot controller

Denso VS-050-S2 @ 2014

- Hydrogen peroxide (H2O2), and ultra violet (UV) resistant
- Hygienic design
- Electro-less Nickel coating

FANUC 200iD @ 2014

- Extreme light weight of 25kg
- Compact &high performance
- Magnesium upper arm

Small robots comparison

hygienic design	dual arm	7-DOF	compact and high performance
_____	_____	_____	_____

human-robot collaborative	Ultra violet resistant	extreme
_____	_____	_____

magnesium

_____

Common Robot

This figure shows the common robots which are now used in industry workshops. Please match the names in the box with the coordinate robots together.

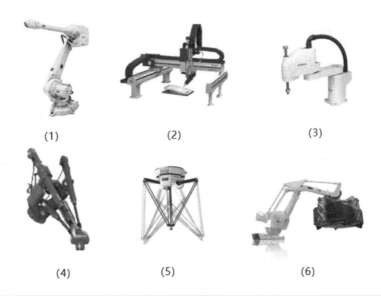

(1)　　　　　　(2)　　　　　　(3)

(4)　　　　　　(5)　　　　　　(6)

articulated robot	delta robot	parallel robot	palletizing robot
_____	_____	_____	_____

gantry robot	selective compliance assembly robot arm
_____	_____

A **delta robot** is a type of parallel robots.It consists of three arms connected to universal joints at the base. The key design feature is the use of parallelograms in the arms, which maintains the orientation of the end effector. By contrast, a Stewart platform can change the orientation of its end effector.

Delta robots have popular usage in picking and packaging in factories because they can be quite fast, some executing up to 300 picks per minute.

What is Industry 4.0? Please read the text and answer the questions.

The concept of cyber-physical systems (CPS) was first defined by Dr. James Truchard, CEO of National Instruments, in 2006, based on a virtual representation of a manufacturing process in software. In January 2012, the German Federal Ministry of Education and Research set up a working group to draft comprehensive strategic recommendations for implementing "Industry 4.0", a term coined by the group. The Industry 4.0 Project is now part of the German government's official High-Tech Strategy, which is actively pursuing in conjunction with private sector partners. Discussions about Industry 4.0 took center stage at April's Hannover Fair, which is why we are suddenly hearing about it.

Industry 4.0 is currently more of a vision than a reality, but it is one with potentially far reaching consequences; and the concept continues to evolve as people think of innovative ways to implement it. However, the following things are already clear.

Sensors will be involved at every stage of the manufacturing process, providing the raw data as well as the feedback that is required by control systems.

Industrial control systems will become far more complex and widely distributed, enabling flexible, fine-grained process control.

RF technologies will tie together the distributed control modules in wireless mesh networks, enabling systems to be reconfigured on the fly in a way that is not possible with hard-wired, centralized control systems.

Programmable logic will become increasingly important since it will be impossible to anticipate all the environmental changes to which control systems will need to dynamically respond.

Smart, connected embedded devices will be everywhere, and designing and programming them will become much more challenging — not interesting and rewarding.

Most of the techniques and technologies needed to implement Industry 4.0 exist today. For example, the radios, sensors, and GPS modules used for asset tracking could just as easily track circuit board assets around the factory floor as they evolve from slabs of FR4 into server blades. The Industry 4.0 spin is that instead of simply attaching an RFID tag and passively tracking the PCB down a linear assembly line, the pick-and-place module could alert inventory when it was running short of memory chips. If the response was that they could not be restocked in time, then all the relevant machines in the entire factory — from the cutting machines and drill presses right through to the systems assembly robots — would reprogram themselves to begin producing the next product for which all parts were in stock, drawing them down from remote inventory as needed, automatically delivered to the right machine just-in-time. Meanwhile, second-source suppliers would be alerted and their assets automatically reconfigured accordingly. The result would be an enormous savings in time and cost versus what even current heavily automated factories can deliver.

Source: https://www.mouser.cn/applications/industry-40/

Questions

1. When and who first mentioned "Industry 4.0"?

_____

2. What is "Industry 4.0"?

_____

_____

_____

_____

3. There are many technology methods used in Industry 4.0. Please complete the sentences.

(1) The sensor will be _____

_____

(2) The programmable logic controller will be _____

_____

(3) Industrial control system will be _____

_____

(4) RF technology will be _____

_____

Translation

If the response was that they could not be restocked in time, then all the relevant machines in the entire factory — from the cutting machines and drill presses right through to the systems assembly robots — would reprogram themselves to begin producing the next product for which all parts were in stock, drawing them down from remote inventory as needed, automatically delivered to the right machine just-in-time.

_____

_____

_____

请和教师交流结果。

教师签名: _____

2.13 Transformer (1)

Task Content: Peter Karl is now busy with a power station product. Please help him choose and buy the transformer. You must know the parameters of transformer and the common problems of a transformer.

Please read the transformer parameters.

Transformer Type		Three-phase Electric Furnace Transformer		
Type of Installation		Indoor		
• Rated Power		A_n	50 +20% 60 Owerload	MVA
❖ Rated Voltage	Primary Voltage	V_{1n}	35	kV
	Secondary Voltage	V_{2n}	0.750+0.442	kV
❖ Rated Current	Primary Current	I_{1n}	825	A
	Secondary Current	I_{2n}	47700 max	A
❖ Frequency		f	50	Hz
❖ Cooling System		OFWF		
❖ Connection Symbol		H.V. ⇒ Delta	Dd	
		L.V. ⇒ Delta closed		

Please fill in the blanks. Try your best and don't use dictionary.

变压器类型	
安装方式	
额定容量	
额定一次侧电压	额定一次侧电流
额定二次侧电压	额定二次侧电流
工作频率	冷却方式
一次侧连接方式	二次侧连接方式

Read the text and translate the steps to assemble transformer's bushings.

INSULATION BUSHINGS

Bushings are usually delivery fully assembled ready to be fitted on the transformer.

➢ **To assemble**

I. Put the flat gaskets and ring nut on the tank cover (6).

II. Connect the connections inside the bushing attachment pin (9).

III. Place the porcelain (8) in the hole on the cover. Make sure the gasket is in a central position.

IV. Lift the ring nut (6) so that the fixing devices (7) can be slipped onto the stud bolts and porcelain. Secure the flange (6) with the nuts and washers (5).

V. Tighten the nuts alternatively and uniformly.

VI.. Place the gaskets on the top part of the porcelain, securing the ring nut (4) and cap (3). Tighten the nut of the through bolt (2).

VII. After filling the transformer with oil, bleed air from the bushing with the bleed screw (4) on the top of the bushing.

During this operation, the bleed screw shall be loosened until oil starts to come out.

1. 本步骤为变压器线套管的_____。

2. 通常线套管在安装前已经_____。

3. 安装步骤

① _____

② _____

③ _____

④ _____

⑤ _____

⑥ _____

⑦ _____

⑧ _____

在整个操作过程中，必须松开排气螺栓，直到热导油可以顺利流出。

Translate the new words or phrases in the word box into Chinese.

transformer bushing	flat gasket	tank	attachment pin
_____	_____	_____	_____
porcelain　　nut	stud bolt	cap	bleed screw
_____　_____	_____	_____	_____

Dialogue

May Lin: Peter, what are you busy with? Is there anything I can help?

Peter: I want to choose some accessories for the new product.

May Lin: You mean the power transformer product?

Peter: Yes, I am now choosing the small accessories for that.

May Lin: These screws are used in transformers. I learned the dimensions of the screws. I know all the screws are made under standard.

Peter: Yes. That is as what you said.

May Lin: But why the nuts used on transformers need current parameters? These nuts need to be connected into the power supply system?

Peter: That is not used on transformers. It is used on the central conductors.

May Lin: What is a central a conductor nut? I know that central conductor is wire.

Peter: A central conductor nut is a component which can protect the power supply system. Usually they are fixed on the bushings, connecting with wires and bushings. But that part is not necessary. That means you also can buy a transformer without a central conductor nut, and that kind of transformers will be cheaper.

May Lin: Oh, I see.

	Current A	Dimensions	Torque Nm
for screws on the stud bolts	-	M10	25
for nuts on the central conductor	1000	M30x2	70
	2000	M42x3	110
	3150	M48x3	180
for the screws of the clamps		M10	25
		M12	40
		M16	90

 Read the dialogue and answer the following questions.

1. What part of a transformer are these accessories used on?

_____

2. What is the central conductor?

_____

3. Can you read the parameters of a screw? What do dimension and torque mean?

_____

 Extend Reading

INDICATIONS

• Do not open the transformer unless environmental conditions are suitable,e.g. no possibility of rain, fog, strong wind, dust, etc.

• Before removing the manholes covers,blow away any dirt which has deposited below the edges. Place a waterproof cover above the transformer.

• The operator shall secure work tools before using them inside the tank.

• Foreign matter inside the tank may be extremely harmful for the safety of the unit.

• Check if the core is grounded, as soon as possible.

• This connection is near a flange on the cover and is made of a copper plait connecting the core to the cover.

• Alternatively, the core is connected to a small bushing(1kV), fitted on the transformer cover, and through this connected to earth (see Overall dimensions drawing).

• Make sure the core has no accidental earthing points, disconnecting the connection from the frames connection, and measuring with a Megger the insulation resistance between the small bushing head and the disconnected connection.

　　1. 上文的主要内容大体围绕什么主题？
　　2. 这些条款的主要对象是哪些群体？
　　3. 上文中反复强调的内容有哪些？

 请和教师交流结果。

教师签名：_____

2.14　Transformer (2)

　　TASK CONTENT: Peter Karl is now busy with a power station product. Please help him choose and buy the transformer. You must know the parameters of transformer and the common problems of a transformer.

Please read the common troubleshooting of a transformer.

- **OIL LEVEL INDICATOR**

Problem	Probable cause	Actions/Possible remedies
Alarm or trip enabled.	The oil level is too high or too low.	Adjust the oil level in relation to ambient temperature, if the transformer is not in service, or to the temperature measured with the "maximum oil thermometer" if the transformer is in service. Check if there are any oil leaks and remedies.

- **ELECTRIC FANS**

Problem	Probable cause	Actions/Possible remedies
The electric fan will not start.	I. No power or a power drop. II. Automatic switch and/or fuses enabled. III. Overload protection enabled.	I. Ensure the right power value. II. Reset the switch and/or replace the fuses. III. Reset the protection.
The electric fan vibrates and is noisy.	I. Wrong rotation direction. II. Worn bearings. III. Rotor not balance.	I. Make sure the rating plate voltage and frequency match that of the power line.Remove the terminal board cover, loosening the screws and checking if connections have been made as indicated inside the terminal board cover. II. Replaced the bearings. III. Disassemble and rebalance the rotor.

Read the common troubles hooting on a transformer, and fill in the blanks.

油位指示器

问题	故障现象	解决方法
	油位过高或者过低	

风扇

问题	故障现象	解决方法
风扇不能启动 风扇有振动且噪音很大		1.确定正确连接电源 2.检查保险丝

Read the accessories of a transformer. Please fill in the blanks.

Accessories Accessorio	Fabricant Fornitore	Type Tipoy-sigla	Drawing reference Riferimento disegno	
1. H.V. BUSHING	BARBERI	52 kV 1000A DIN 42534	TAD010000000402	1
2. CORE EARTHING BUSHING	BARBERI	1 kV 630A-DIN 42530	TAD010000000402	24
3. ON LOAD TAP CHANGER	MASCHINENFABRIK REINHAUSEN	M III 600D 72.5/B16 150	TAD010000000402	3
4. O.L.T.C MOTOR DRIVE UNIT	MASCHINENFABRIK REINHAUSEN	ED L 100	TAD010000000402	4
5. OFF CIRCUIT TAP CHANGER	CAPT	KLD 360 36kV 1200A n° 6 tap MOTOR DARIVER 070-1.12.310	TAD010000000402	5
6. OIL WATER COOLING UNIT	GEA	WKDH-800 Eco/Z Double tube	TAD010000000402	28
7. OIL MOTOR PUMP COOLING UNIT	GEA	W4/210/125	TAD010000000402	29
8. OIL FLOW INDICATOR	GEA	FLOW SENSOR SI 5200 CONTROL MONITOR VS 3000	TAD010000000402	29
9. WATER FLOW INDICATOR	IFM	CCL 2 O	TAD010000000402	29
10. OIL LEVEL INDICATOR (TRANSFORMER)	ETI	IMLO R 220/F/M	TAD010000000402	13
11. OIL LEVEL INDICATOR (OLTC)	CEDASPE	IMLO R 220/F/M	TAD010000000402	14
12. DEHYDARTING SILICAGEL AIR BREATHER	CEDASPE	EIF 742	TAD010000000402	33

1. 这是一张什么表格？
2. 表格中主要提供了哪些信息？

配件名称	型 号
高压线套管	
油浸冷却单元	
	W4/210/125
接地套管	
	IMLO R 220/F/M
O.L.T.C 电机驱动单元	

Translate the new words or phrases in the word box into Chinese.

ambient temperature	maximum oil thermometer	trip
_____	_____	_____
vibrate and noisy	overload protection	automatic switch
_____	_____	_____
oil level indicator	flow sensor	bearing
_____	_____	_____

The following diagram is part of a real electrical circuit drawing.

Li Jun now is translating this drawing. Please help him.

导线型号： _____

工作温度： _____

温度传感器型号： _____

安装要求： _____

_____

电压等级： _____

导线用途及颜色：红黄绿/黑＿＿＿＿＿＿＿＿＿＿＿＿＿＿＿＿＿＿＿

　　　　　　　黄/绿＿＿＿＿＿＿＿＿＿＿＿＿＿＿＿＿＿＿＿＿

　　　　　　　红＿＿＿＿＿＿＿＿＿＿＿＿＿＿＿＿＿＿＿＿＿＿

Wires Type: N07V-K

Normal Voltage Vo/V: 450/750V

Test Voltage: 2500V a c

Working Temp. : -20℃~90℃

For Temperature Sensor Cable Type: FROH2R 450/750V (3×1+screen)

Interconnection from control box and transformer devices is done by single wires inserted in galvanized carbon steel

Flexible conduit with single seam and PVC covering.

VOTAGE LEVEL:　X1　THREE - PHASE POWER SUPPLY 380V/3ph~50Hz

　　　　　　　　　　SINGLE - PHASE POWER SUPPLY -CONTROL CIRCUIT

　　　　　　　X2　TRANSFORMER DEVICES PROTECTION CONTACTS

　　　　　　　X3　CURRENT TRANSFORMERS

VLOTAGE LEVEL :　　THREE - PHASE CURRENT 380V/3ph~50Hz　RED YELLOW GREEN/BLACK

　　　　　　　　　PROTECTION　　　　　　　　　　　　GREEN/YELLOW

　　　　　　　　　AUXILIARY VOLTAGE 220 Vac　　　　RED

Don't use a dictionary. Test yourself. Can you understand?

Problem	Probable cause	Actions / Possible remedies
Turning on the three-phase power mains sectionalising switch the signal lamp does not work	Bad power supply on three-phase input	Check the power mains of three-phase power supply

这是一个什么故障？如何解决这个故障？

 请和教师交流结果。

教师签名：＿＿＿＿＿＿＿

2.15   Measuring Instrument

Task Content: Once a particular component is suspected of being faulty, individual tests must be then performed. And the testing and measuring instruments often to be used are as follow.

Please read the text, and answer the questions.

Multimeter

Using a multimeter, you should first place it flat on a table and note if the meter pointer indicates exactly 0 at the extreme left end of the black scale. If it doesn't read 0, turn the screw on the meter movement slowly until the proper 0 reading is realized. And then connect test-lead to the other．Set the RANGE switch to the proper item and range．Insert the test-lead plugs into the correct pair of jacks, clip or hold the test probes on the terminals of the part being checked, and read the meter.

A multimeter is sometimes called a volt-ohm—milliammeter. Multimeters having only volt and ohm ranges are also called volt-ohmmeters. There are two main kinds of multimeters. They are mechanical handy multimeters and digital ones. It's better for learners to use handy types, because it's helpful for them to be familiar with some electronic principles.

With a multimeter, we can measure current, voltage, resistance, etc. And we can also troubleshoot appliances, test fuses, measure voltage at a wall outlet, check the wires from your roof antenna, find out whether a wall switch is broken, and perform dozens of other household jobs. In troubleshooting, voltage can be measured while the set is on, and resistance can be measured while the set is off. You get clues to the location of the trouble by comparing your multimeter readings with the normal values given in service manuals.

Questions

1. What do you do first when you use a multimeter?

_____

_____

_____

2. What can a multimeter measure?

_____

_____

_____

3. What is the difference between volt-ohmmeters and multimeters?

_____

_____

_____

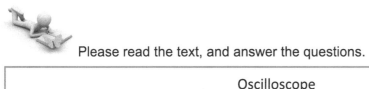 Please read the text, and answer the questions.

Oscilloscope

An oscilloscope is a display instrument. With it you can study the wave-form of an electric signal, such as the wave-form of alternating current and voltage. You can measure voltage, current, power, and frequency as well. In fact, almost any quantity that involves amplitude and wave-form.

The heart of the oscilloscope is the cathode-ray tube (CRT). This consists of the base, neck (an electron gun is included), bulb and the face-plate (screen). The electron gun consists of a cathode, a control grid, an anode and two sets of deflection plates. When the cathode is heated, it emits electrons, which form electron beam. The electron beam sweeps rapidly from left to right across the screen of a cathode-ray tube. However, recently most of the cathode-ray tubes used in oscilloscopes have been replaced by LCDs (Liquid Crystal Displays), which greatly decrease the weight and size of the instrument.

Questions

1. What type of the instrument is an oscilloscope?

_____

_____

2. What is the heart of an oscilloscope?

_____

_____

3. Which parameter is decreased when the LCDs replace the cathode-ray tubes?

_____

_____

Let's do some exercises.

I. Choose the best answer to complete the following statements.

1. According to the text, _____is the most common tool in repairing work.
    A. an oscilloscope   B. a multimeter
    C. a screw   D. an electron gun

2. The multimeter must be put horizontally and the pointer must point to_____
    A. the left   B. the end
    C. zero   D. range

3. Different _____ must be set for different testing and measuring.
    A. screws   B. range switches
    C. probes   D. plugs

4. When testing or measuring you should hold the test probes on the terminals of the _____.
    A. parts being checked   B. meters
    C. jacks   D. plugs

5. An oscilloscope is an instrument which can_____ the wave-form of an electric signal.
    A. study   B. draw
    C. display   D. generate

6. _____greatly decrease the weight and size of the oscilloscopes.
    A. LCDs   B. CRTs
    C. Screens   D. Electron guns

II. Translate the following phrases into English.

1．机械指针式万用表

_____

2．阴极射线管

_____

3．液晶显示器

_____

4．电信号

_____

5．两套偏转板

_____

6．电子原理

_____

请和教师交流结果。

教师签名：_____

微课资源

扫一扫：获取相关微课视频。

1.2 My workshop

2.1　Electric Circuit
　　(1)-1-1

2.1　Electric Circuit
　　(1)-1-2

2.1　Electric Circuit
　　(1)-1-3

2.1　Electric Circuit
　　(1)-1-4

2.1　Electric Circuit
　　(1)-2

2.1　Electric Circuit
　　(1)-3-1

2.1　Electric Circuit
　　(1)-3-2

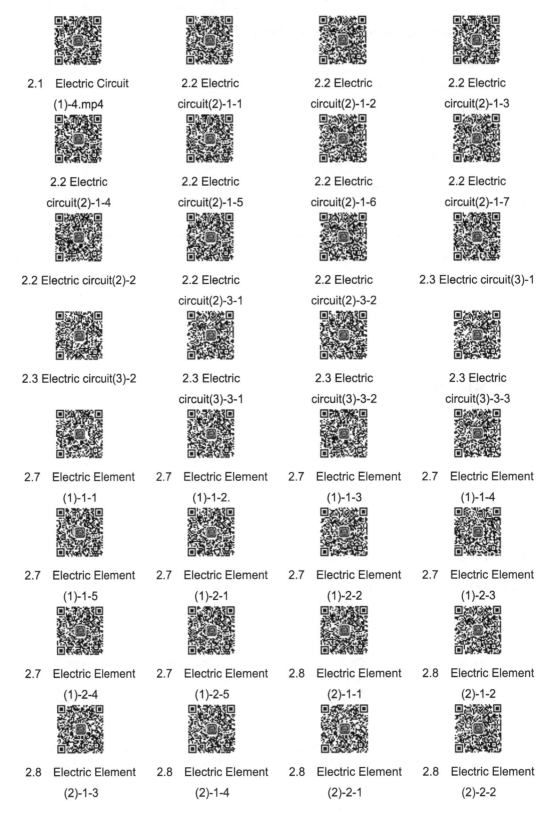

2.1 Electric Circuit (1)-4.mp4
2.2 Electric circuit(2)-1-1
2.2 Electric circuit(2)-1-2
2.2 Electric circuit(2)-1-3
2.2 Electric circuit(2)-1-4
2.2 Electric circuit(2)-1-5
2.2 Electric circuit(2)-1-6
2.2 Electric circuit(2)-1-7
2.2 Electric circuit(2)-2
2.2 Electric circuit(2)-3-1
2.2 Electric circuit(2)-3-2
2.3 Electric circuit(3)-1
2.3 Electric circuit(3)-2
2.3 Electric circuit(3)-3-1
2.3 Electric circuit(3)-3-2
2.3 Electric circuit(3)-3-3
2.7 Electric Element (1)-1-1
2.7 Electric Element (1)-1-2.
2.7 Electric Element (1)-1-3
2.7 Electric Element (1)-1-4
2.7 Electric Element (1)-1-5
2.7 Electric Element (1)-2-1
2.7 Electric Element (1)-2-2
2.7 Electric Element (1)-2-3
2.7 Electric Element (1)-2-4
2.7 Electric Element (1)-2-5
2.8 Electric Element (2)-1-1
2.8 Electric Element (2)-1-2
2.8 Electric Element (2)-1-3
2.8 Electric Element (2)-1-4
2.8 Electric Element (2)-2-1
2.8 Electric Element (2)-2-2

2.8　Electric Element
(2)-2-3

2.9　Electric Element
(3)-1-1

2.9　Electric Element
(3)-1-2

2.9　Electric Element
(3)-1-3

2.9　Electric Element
(3)-1-4

2.9　Electric Element
(3)-1-5

2.9　Electric Element
(3)-1-6

2.9　Electric Element
(3)-1-7

2.9　Electric Element
(3)-1-8

2.9　Electric Element
(3)-1-9

Module 3　Modern Manufacture Technology

3.1　3D Printing

Task Content: What is a 3D printer? What kind of materials does a 3D printer use instead of ink? What problem might 3D printing cause?

A 3D Printer

Imagine having an idea, drawing it on paper, bringing it to a store and seeing it turned into a physical object? This is now possible with the help of 3D printers. A 3D printer just as a traditional printer sprays ink onto paper line by line, modern 3D printers spread materials onto a surface layer by layer, from the bottom to the top, gradually building up a shape.

Instead of ink, the materials of 3D printer uses are mainly plastic, resin and certain metals. The thinner each layer is, from a millimeter to less than the width of a hair, the smoother and finer the object will be. 3D printers were once used just by universities and big companies, but now, stores with 3D printing services are appearing around the United States. 3D printing services are becoming available for American consumers. The UPS Store is a nationwide retailer that provides 3D printing services.

However, as 3D printing becomes more commonplace, it may trigger piracy. "Once you download a coffee maker, or print out a new set of kitchen utensils on your personal 3D printer, who will visit a retail store again? " An expert in 3D printing told Forbes News. Even more frightening, what if someone in the world uses a 3D printer to print out a fully functioning gun?

Match Exercises

A. convenient in maintenance	1. (　　)质量可靠
B. above money's worth	2. (　　)物超所值
C. hold a patent right	3. (　　)享有专利
D. easily assembled	4. (　　)装配方便
E. innovative design	5. (　　)结构坚固
F. dependable quality	6. (　　)安全性能好
G. high safety	7. (　　)维修方便
H. firm in structure	8. (　　)创新设计

Here are 10 cool and practical 3D prints for offices. Can you figure them out?

If you're any one like me, spending a lot of time in your office trying to keep productive, sometimes you can become complacent in your environment and forget about optimizing it to make your space as functional, inspiring and efficient as possible.

1.

2.

3.

4.
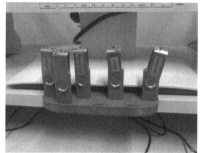

5.

6.

7.

8.

9.
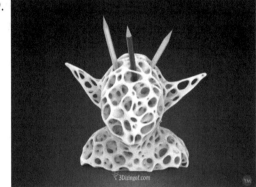

10.

A USB Rack: (　　)

USB sticks are like public toilets, when you want one you can never find one, and when you do it's filled with crap. Get your USB organized and place an order for this functional USB rack by Makulator.

A Keyboard Holder: (　　)

If you're looking to optimised space, then a keyboard holder could be a great idea. If you can stow away your keyboard with ease, it will free up your workspace for other productive tasks. Credit to Erik Cederberg for this space saving design.

Headphone Holder: (   )

Similar to the previous object, this headphone holder by Nadar will help to keep your desk free from unnecessary clutter. Have you ever had your headphones fall on the ground only to get caught around you chair legs? A holder will help that happen less, it will still happen, but less often.

Volume Voronoi Yoda: (   )

Put your pencils away in style with a cool pencil holder! There are various 3D printed pencil holder models but my favourite one is this "Volume Voronoi" Yoda holder by Asher Nahmias.

Low Poly Extinction: (   )

If you're a little more minimalist and you like to keep fewer pens around, why not stick it in this dying T-rex's mouth. Thanks to XYZ Workshop for this design.

SD Card Organiser: (   )

This one is pretty self-explanatory. If you're like me and have SD cards flopping about in a drawer, then you should print one of these SD card organisers. This design is by Neon Green.

A Coffee Table: (   )

Why do a "coffee table" have to be so big anyway? It's just for your coffee right? This has to be the world's first properly-sized coffee table. Perfect for keeping your hot coffee off your desk or notepad. You could even have a tiny lunch on it if you're on a diet. Designed by Creative Tools.

Tape Dispenser: (   )

Talking about practical 3D prints: Kick-start your office supply with this unique looking tape holder designed by Ysoft.

Desktop Aquaponics System: (   )

Having some nature around the office can really help to improve your mood. If you'd like to have something different, why not experiment with your own desktop aquaponics system?

Office Basketball Set: (   )

Take a break from the keyboard with this office basketball set designed by Ultimaker. It comes with various bracket sizes, a screw bracket and a ball, everything you need to shoot some hoops.

 Laminated Object Manufacturing (LOM)

Look at the picture. Please match each part's number of LOM with the given words or phrases in the box.

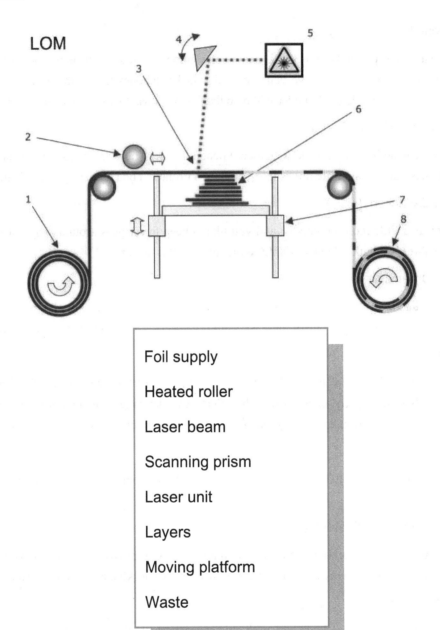

Video Time:

Can you really print out edible Oreos? CNN's Laurie Segall taste tests a 3D-printed Oreo at South by Southwest. Here is their dialogue.

—So first of all, tell me a little bit about what we are looking at here.

—Of course, so you are looking at the trending/vending machine from Oreo. So the whole idea here was how we begin to explore customizing flavors for consumers, and really connecting that consumer experience to technology in a way that has never been done before.

—A part of the technology behind here is 3D printing.

—Yes, so we use 3D printing parts and 3D printing approaches.

—How does this transit to other types of food? Could we 3D print tacos one day, perhaps?

—I think one day, you are able to 3D print a lot of things, everything from chocolate to candle, but again, I think it's not so much just about technology, although it's a breakthrough, it's really about how we begin to understand what consumers want and to liberalize those kinds of customized experiences.

—At what point are we going to be able to 3D print our own customizable Oreos from our home?

—That's a good question. I don't have a direct timeline, but we would love to be able to deliver that customized experience, so consumers could have all flavors that we offer as well as create their own.

—Already.

—OK, here we go.

—OK. All right. It looks good. It looks like a normal one, just OK.

—Yes, go ahead.

—Really good.

—See, now you can just go over to the milk bar and dunk it in the milk.

—Yeah, that is really good, taste like an Oreo.

—Here you go.

Talking

After watching the video, what kinds of 3D products do you want to make?

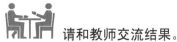 请和教师交流结果。

教师签名：＿＿＿＿＿＿＿

3.2 Engineering Drawing

Task Content: Peter Karl is preparing the drawing instruments and he will make a detail drawing. Do you know basic knowledge on mechanical engineering drawing?

Match the given pictures with the words or phrases.

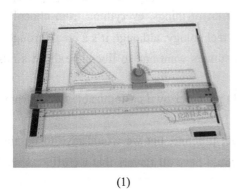

(1)

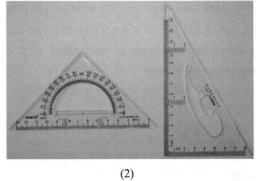

(2)

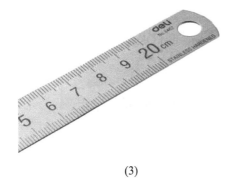

(3)

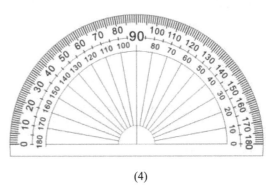

(4)

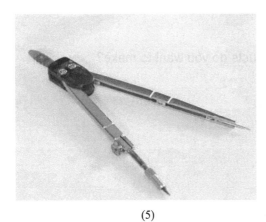

(5)

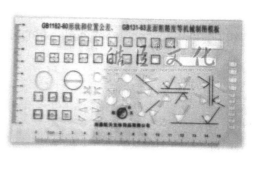

(6)

Drawing board   Protractor

Template   Steel rule

Set square   Compasses

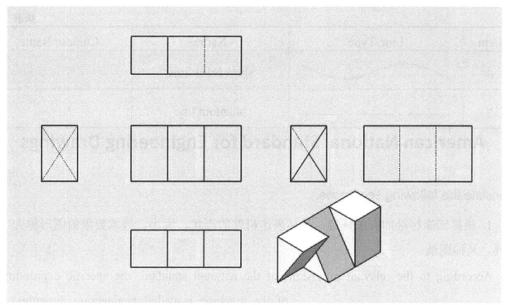

Developed Representation of the Six Views

Speak out the names of the views.

left-side view

front view

bottom view

vertical view

right-side view

rear view

Translate the names of the following line types into Chinese.

Item	Line Type	Name	Chinese Name
1	————————	visible line	
2	— — — — — — — —	hidden line	
3	————————	section line	
4	——— – ——— – ———	center line	
5	——∿—∿——	long break line	

续表

Item	Line Type	Name	Chinese Name
6		short break line	
7		phantom line	

American National Standard for Engineering Drawings

Translate the following sentences.

1. 根据国家标准的有关规定，具体表达机件的形状、大小、技术要求的图形称为工程图样，又称图纸。

According to the relevant provisions of the national standard, the specific expression of the _____ , _____ , _____ of the machine is called engineering drawings, also known as drawings.

2. 图形与实物的线性尺寸之比称为比例。一般采用 1∶1 比例。图样中标注的尺寸均为机件的实际加工尺寸。

The ratio of _____ of the figure and _____ is called the proportion, and the proportion of the 1∶1 is usually used. The dimensions in the drawing are _____ dimensions.

3. 投射线互相平行并且垂直于基本投影面的投影法称为正投影法。正投影法是绘制工程图样的基本投影法。

That the rays are _____ to each other and _____ to the projection plane is called the projection method. Orthographic projection is a basic projection method for drawing engineering drawings.

4. 空间物体共有长、宽、高三个方向，分别反映左右、前后、上下六个方位。每个视图分别对应两个方位。

Space objects have a _____ , _____ , _____ three directions, respectively, reflecting the _____ , _____ , _____ six directions.Each view corresponds to two directions.

 Questions for Text

1. What is a coordinate system composed of?

2. What is called a projection?

3. Why do we call engineering drawing an abstract universal language?

Engineering Drawing

The basic of all input Auto CAD is the Cartesian coordinate system, and the various methods of input (absolute or relative) rely on this system.

The fixed Cartesian coordinate system locates all points on an Auto CAD drawing by defining a series of positive and negative axes to locate positions in space. There is a permanent origin point(0,0) which is referenced, an X axis running horizontally in a positive and negative direction from the origin, and a Y axis travelling per-pendicularly in a vertical direction. When a point is located, it is based on the origin point unless you are working in three dimensions, in which case, you will have a third axis, called the Z axis.

Projection

An orthographic projection of an object is seen from the front, top, right side, etc.

Engineering drawing is an abstract universal language used to represent a designer's idea to others. It is the most acceptable medium of communication in all phases of industrial and engineering work.

Any object can be viewed from six mutually perpendicular directions. These six views may be drawn if necessary.

A drawing for use in production should contain only those views needed for a clear and complete shape description of the object. These minimum required views are referred to as necessary views. In selecting views, the drafter should choose those that best show essential contours or shapes and have the least number of hidden lines.

The three-principle dimensions of an object are width, height, and depth. The top, front, and right-side views are arranged closed together. These are called three regular views because they are views used most frequently.

 请和教师交流结果。

教师签名：＿＿＿＿＿＿＿＿＿

3.3　Tolerances and Fits

Task Content: Peter Karl now is trying to read a technical drawing. And some technical requirements need learning. Please learn with him.

Put the technical terms into the frames.

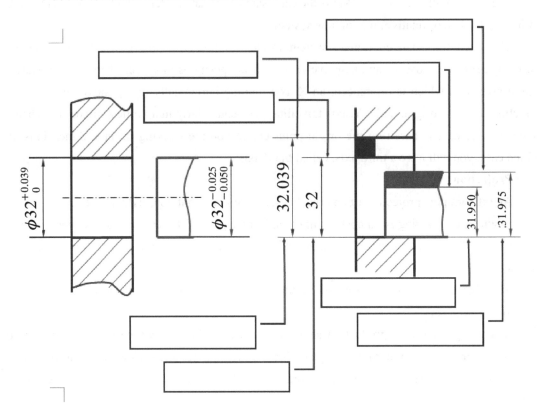

Upper deviation   Lower deviation
Maximum limit of size   Minimum limit of size

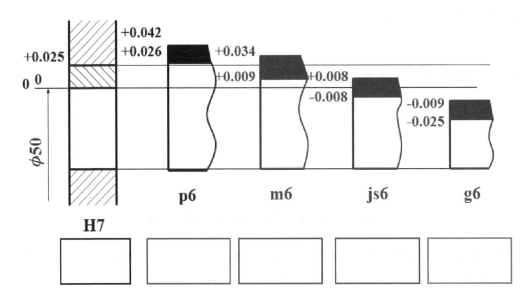

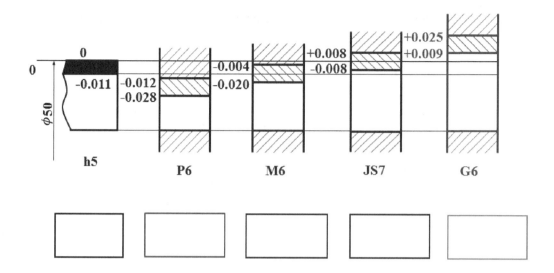

<table>
<tr><td></td><td></td><td></td><td></td><td></td></tr>
</table>

Hole basis fit

Shaft basis fit

Interference fit

Transition fit

Clearance fit

Translate the symbol names into Chinese.

Item	ANSI Y14.5M	Symbol for	Chinese Name
1	▬	straightness	
2	▱	flatness	
3	◯	roundness/circularity	

Item	ANSI Y14.5M	Symbol for	Chinese Name
4		profile of a line	
5		profile of a surface	
6		parallelism	
7		perpendicularity	
8		angularity	
9		symmetry	
10		concentricity/coaxiality	
11		circular runout	
12		total runout	
13		position	

Read the following drawing and answer some oral questions.

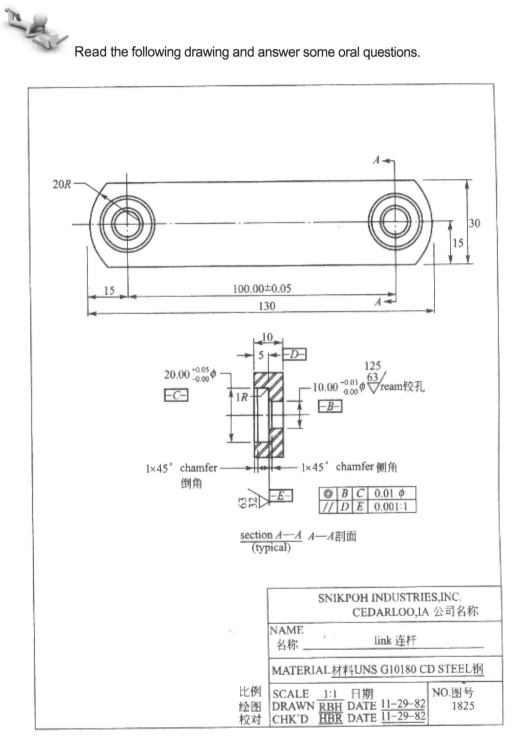

An example of a detail drawing 零件图例

Reading Comprehension

 Questions for Text Discussion

1. Why is it impossible to make anything to the exact size?
2. What is tolerance?
3. What are three classes of fits?

Tolerances and Fits

Interchangeable manufacturing allows parts made in widely separated locations to be brought together for the end assembly. That the parts all fit together properly is an essential element of mass production. Without interchangeable manufacturing, modern industry could not exist, and without effective size control by the engineers, interchangeable manufacturing could not be achieved.

However, it is impossible to make anything to the exact size. Parts can be made to very close dimensions, even to a few millionths of an inch or thousandths of a millimeter, but such accuracy is extremely expensive.

Fortunately, exact sizes are not needed. The need is for varying degrees of accuracy according to functional requirements. So what is wanted is a means of specifying dimensions with whatever degree of accuracy is required. The answer to the problem is the specification of a tolerance on each dimension.

The tolerance is the total amount that a specific dimension is permitted to vary; it is the difference between the maximum and minimum limits for the dimension. In engineering when a product is designed, it consists of a number of parts and these parts mate with each other in some form. In the assembly it is important to consider the type of mating or fit between two parts which will actually define the way the parts are to behave during the working of the assembly.

The fit between two mating parts is the relationship which results from the clearance or interference obtained. There are three classes of fits, namely, clearance, transition and interference.

 请和教师交流结果。

教师签名：_____

3.4   Mechanical Design

Task Content: Do you want to be a mechanical designer? Are you a qualified one? Receive the test and see the result.

Match the given mechanisms with the given words or phrase.

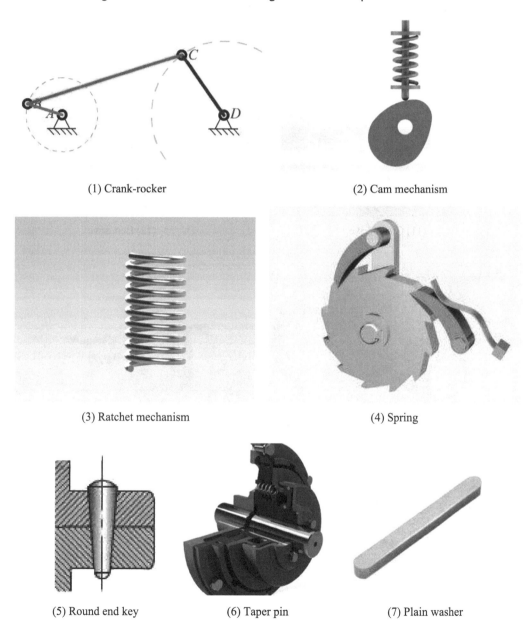

(1) Crank-rocker

(2) Cam mechanism

(3) Ratchet mechanism

(4) Spring

(5) Round end key

(6) Taper pin

(7) Plain washer

(8) Bolt coupling

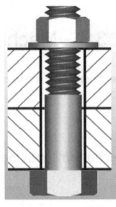

(9) Screw coupling

(10) Screw drive

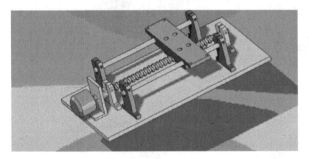

(11) Chain drive

(12) Belt drive

(13) Gear drive

(14) Ball bearing

(15) Shaft

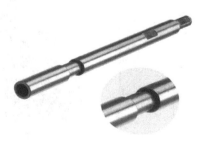

(16) Friction clutch

Do you know the following famous mechanical inventions and their inventors?

_____

_____

_____

_____

Reading Comprehension

 Questions for Text Discussion

1. What does Karl major in?
2. Why did Karl choose this major?
3. Why is Karl primarily interested in mechanical engineering?
4. Why did Karl decide to apply for this position?
5. What will determine an employee's progress in a company in Karl's opinion?

Are you a qualified mechanical designer?

(Karl: Applicant   Mrs. Smith: Interviewer)

Mrs. Smith: Come in, please.

Karl: Good afternoon, Mrs. Smith.

Mrs. Smith: Good afternoon. Have a seat, please. Are you Mr. Brown?

Karl: Thank you. Yes, I am Karl Brown.

Mrs. Smith: I have read your resume. I know you have worked for 3 years. Why did you choose to major in mechanical engineering?

Karl: Many factors led me to choose mechanical engineering. The most important factor is that I like tinkering with machines.

Mrs. Smith: Why are you primarily interested in mechanical engineering?

Karl: I like designing products, and one of my designs received an award. Moreover, I am familiar with CAD. I can do any mechanic well if I am employed.

Mrs. Smith: Why did you decide to apply for this position?

Karl: Your company has a very good reputation, and I am very interested in the field your company is in.

Mrs. Smith: In your opinion, what will determine an employee's progress in a company such as ours?

Karl: Interpersonal and technical skills.

Mrs. Smith: OK. Thank you.

Read the dialogue carefully and then decide whether each of the following statements is True (T) or False (F).

(   ) 1. Karl has no experience in the job.

(   ) 2. Karl is very interested in product designing.

(   ) 3. Karl likes fixing the machines.

(   ) 4. Karl' design has ever received an award.

(   ) 5. Karl thinks interpersonal skill is more important than technical skill.

3.5   Machine Tools

Task Content: Most of the mechanical operations are commonly performed on five basic machine tools.

• The drilling press

• The lathe

• The shaper or planer

• The milling machine

• The grinder

Let's learn them together.

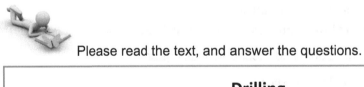

Please read the text, and answer the questions.

Drilling

Drilling is performed with a rotating tool called a drill. Most drilling in metal is done with a twist drill. The machine used for drilling is called a drill press. Operations, such as reaming and tapping, are also classified as drilling. Reaming consists of removing a small amount of metal from a hole already drilled. Tapping is the process of cutting a thread inside a hole so that a cap screw or bolt may be threaded into it.

Questions

1. What is a drilling machine used for?

_____

2. Which operations are drilling?

_____

 Please read the text, and answer the questions.

Turning and Boring

The lathe is commonly called the father of the entire machine tool family. For turning operations, the lathe uses a single-point cutting tool, which removes metal as it travels past the revolving workpiece. Turning operations are required to make many different cylindrical shapes, such as axes, gear blanks, pulleys, and threaded shafts. Boring operations are performed to enlarge, finish, and accurately locate holes.

Questions

1. What is a turning and boring machine used for?

_____

2. Which operations are turning?

_____

3. Which operations are boring?

_____

Please read the text, and answer the questions.

Milling

Milling removes metal with a revolving, multiple cutting-edge tool called a milling cutter. Milling cutters are made in many styles and sizes. Some have as few as two cutting edges and others have 30 or more. Milling can produce flat or angled surfaces, grooves, slots, gear teeth, and other profiles, depending on the shape of the cutters being used.

Questions

1. What is a milling machine used for?

_____

2. Which operations are milling?

_____

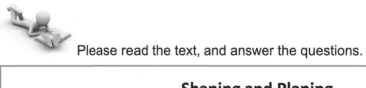

Please read the text, and answer the questions.

Shaping and Planing

Shaping and planing produce flat surfaces with a single-point cutting tool. In shaping, the cutting tool on a shaper reciprocates or moves back and forth while the work is fed automatically towards the tool. In planing, the workpiece is attached to a worktable that reciprocates past the cutting tool. The cutting tool is automatically fed into the workpiece a small amount on each stroke.

Questions

1. What is a shaping and planing machine used for?

_____

_____

2. Which operations are shaping and planing?

_____

_____

 Please read the text, and answer the questions.

Grinding

Grinding makes use of abrasive particles to do the cutting. Grinding operations may be classified as precision or nonprecision, depending on the purpose. Precision grinding is concerned with grinding to close tolerances and very smooth finish. Nonprecision grinding involves the removal of metal where accuracy is not important.

Questions

1. What is a grinding machine used for?

_____

2. Which operations are grinding?

_____

 Let's do some exercises.

I . Answer the following questions according to the text.

1. What parts can be processed with lathes?

_____

2. Which machine tools can process flat surfaces?

_____

3. What is the tapping?

_____

Ⅱ. Decide whether each of the following statements is True(T) or False(F) according to the text.

(1) For turning operations, the lathe uses multiple cutting edge tools to process.

(2) The milling machine is commonly called the father of the entire machine tool family.

(3) In shaping, the workpiece is attached to a worktable that reciprocates past the cutting tool.

(4) Grinding makes use of the metal material tools to do the cutting.

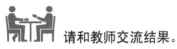 请和教师交流结果。

教师签名：_____

3.6 Introduction of AutoCAD (1)

Task Content: Li Jun now is studying CAD/CAM. He wants to study how to use AutoCAD first.

Translate the words or phrases in the word box into Chinese.

file	new	open	save	save as	export
_____	_____	_____	_____	_____	_____
publish	print	plot preview	close	undo	redo
_____	_____	_____	_____	_____	_____

Reading & Writing

The menu File is very important, including some submenus.

1. **New**: It is used to create a blank drawing file.

2. **Open:** Opens an existing drawing file.

3. **Save:** Saves the current drawing.

4. **Save as:** Saves a copy of the current drawing under a new file name.

5. **Print/plot:** Prints a drawing to a printer, plotter, or other devices.

6. **Export:** Exports to a different format.

7. **Close:** Closes the drawing.

8. **Undo:** Reverses the most recent action.

9. **Redo:** Reverses the effects of the previous Undo or U command.

10. **Plot Preview:** Views the effect of a drawing before printing.

Read the following diagram and filling in the diagram below.

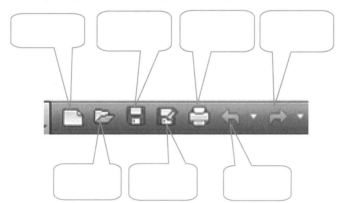

Translate the words in the word box into Chinese.

draw	line	polyline	circle	arc
_____	_____	_____	_____	_____
radius	diameter	tangent	rectangle	polygon
_____	_____	_____	_____	_____
hatch	ellipse			
_____	_____			

Reading & Writing

The menu Draw is very important, including some submenus that you often use.

1. **Line** With Line, you can create series of continuous line segments.

2. **Polyline** Creates a 2D polyline. A 2D polyline is a connected sequence of segments created as a single planar object.

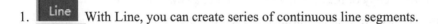

3. **Center, Radius** Creates a circle with a center point and a radius.

Center, Diameter Creates a circle with a center point and a diameter.

2-Point Creates a circle with two points.

3-Point Creates a circle with three points.

Tan, Tan, Radius Creates a circle with two tangent lines and a radius.

Tan, Tan, Tan Creates a circle with three tangent lines.

4. **Arc** Creates an arc with three points.

5. Fills an enclosed area or selected objects with a hatch pattern or fill.

6. Creates an ellipse with a center point, point or axis.

7. **Rectangle** Creates a rectangle from the specified rectangle parameters (length, width, rotation) and type of corners (fillet, chamfer, or square).

8. **Polygon** Creates a polygon by setting number of sides and other parameters.

9. With Helix, you can create a 2D spiral or a 3D spring.

10. With Donut, you can create a filled circle or a wide ring.

Write down the commands when operating.

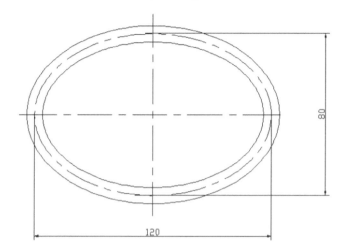

Command 1: _____

Command 2: _____

Command 3: _____

Command 4: _____

Command 5: _____

Command 6: _____

Command 7: _____

Command 8: _____

Write down the commands when operating.

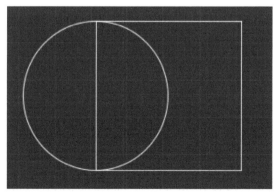

Command 1: _____

Command 2:_____

Command 3:_____

Command 4:_____

Command 5:_____

Command 6:_____

 请和教师交流结果。

教师签名：_____

3.7　Introduction of AutoCAD (2)

Task Content: Li Jun now is studying AutoCAD. He wants to study how to use AutoCAD first.

Translate the words or phrases in the word box into Chinese.

dimension	linear		aligned	angular
_____	_____		_____	_____
radius	diameter	leader	continue	baseline
_____	_____	_____	_____	_____
multiline text	dimension style	modify		
_____	_____	_____		

Reading & Writing

The menu Dimension is very important, including some submenus.

1. ┠─┤ Linear　It is used to create a linear dimension with a horizontal, vertical or rotated

dimension line.

2.  Creates a linear dimension that is aligned with the origin point of the extension line.

3.  Creates an angular dimension. Measures an angle between selected objects or 3 points.

4. Creates a radius dimension for a circle or an arc.

5. Creates a diameter dimension for a circle or an arc.

6. Creates a jogged radius dimension when the center of an arc or circle is located off the layout and cannot be displayed in its true location.

7. Creates a multileader object. A multileader object typically consists of an arrowhead, a horizontal landing, a leader line or curve, a multiline text object or a block.

8. Creates a dimension that starts from an extension line of a previously created dimension.

9. Creates a linear, angular, or ordinate dimension from the baseline of the previous or selected dimension.

10. Creates a multiline text object.

11. Displays texts on screen as it is entered.

机电专业英语项目化教程(微课版)

Write down the procedures of drawing.

Step 1: _____

Step 2: _____

Step 3: _____

Step 4: _____

Step 5: _____

Step 6: _____

Step 7: _____

Step 8: _____

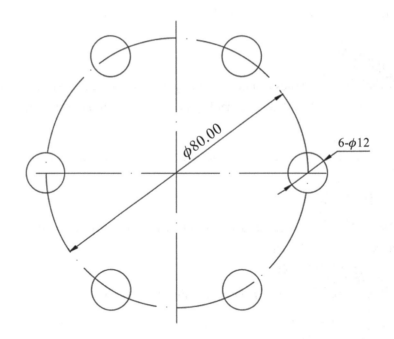

Read the following diagram and introduce it to your partners.

Part A: _____

Part B: _____

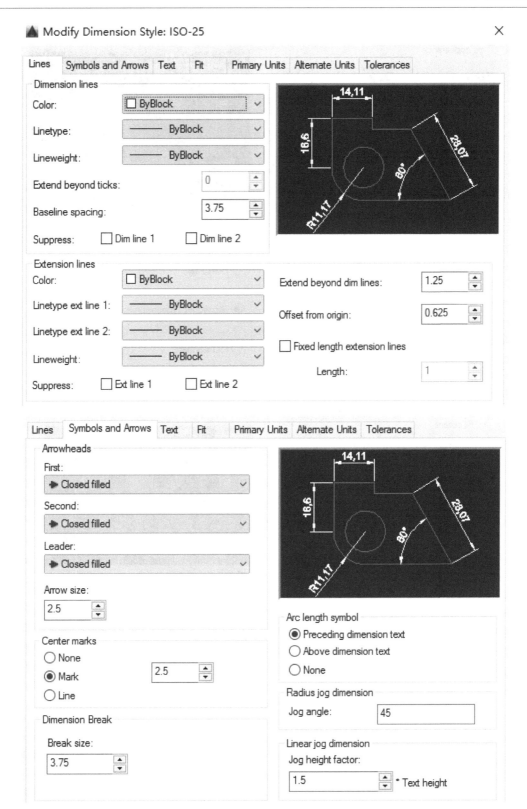

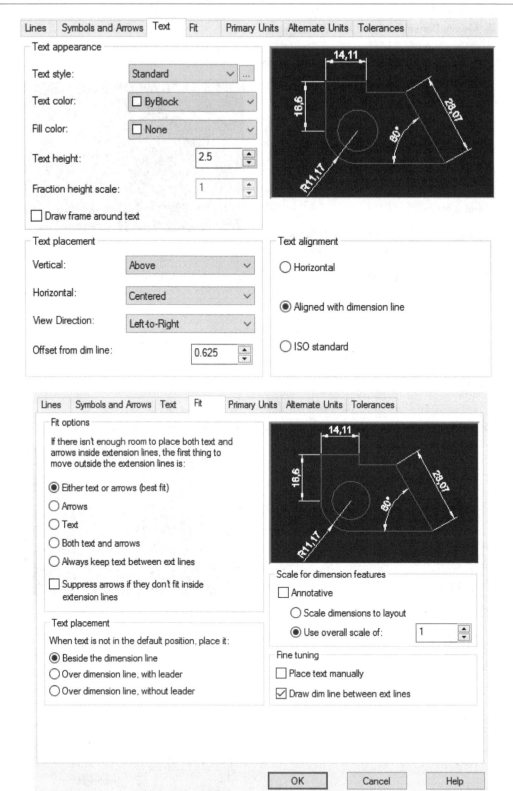

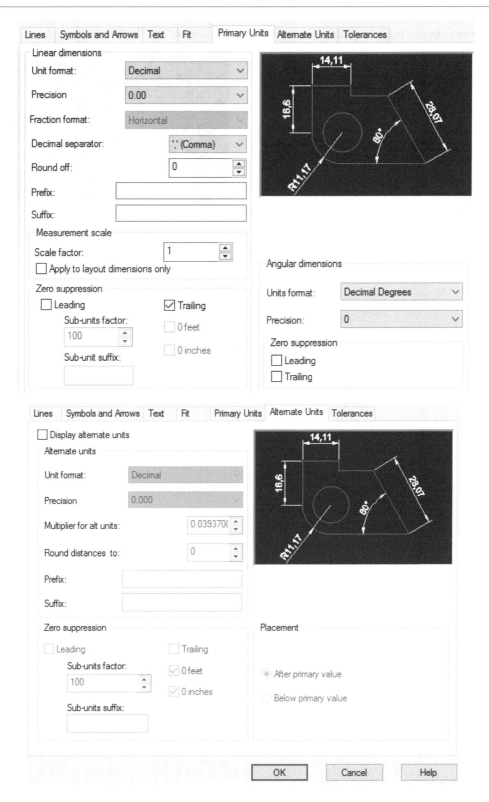

请和教师交流结果。

教师签名：＿＿＿＿＿＿＿＿＿

3.8　Safety and Maintenance

Task Content: Here are some safety notes on the machine in a workshop, please read them and make an induction to your partners.

TASK 1

Recite the signs to your partners.

WARNING

1. Turn off power before performing maintenance.
2. May cause an electric shock.
3. May result in fatal injury or death.

IMPORTANT

1. Must be properly grounded.
2. Electric shock by loose wire.
3. May cause minor injury.

TASK 2

Introduce the CAUTION to your partners.

1. Pay attention to safety during operation. Avoid hazard of the rotary operation unit!

2. Avoid hazard of pulling or drawing the guard door!

3. Avoid hazard of operating/closing the watching door!

4. Avoid hazard of opening/closing the electrical cabinet door!

5. Avoid hazard of opening/closing the pneumatic unit door!

6. Close safety door during machine running, otherwise serious casualty accident may occur!

(1) Wear proper labor protection! Avoid body injuries caused by splashing of fluid or chips!

(2) Protect environment! Avoid chips, fluid splashing or leakage.

(3) Subsequent hurt!

(4) Adopt measures to forbid entering into moving or working range of running machine!

TASK 3

Read General Security Items, recite it to your partners.

1. Read carefully, understand the mechanical and electrical operation manual completely and grasp all safety precautions and proper utilization before operation.

2. Don't approach the motion range of machine during auto running. If the motion range has to be approached, power must be cut off or the machine must stop running to ensure safety status.

3. The machine provides several safety devices for operator's security and protecting machine. Don't remove safety devices to run.

4. Firmly fix workpieces and tools before machining. Don't put any implements on the movable part. Select proper machining parameters for machining.

5. Wear proper security protection things(i.e. clothes, protection glasses, etc). Work safely.

6. Install and service the machine by professionals according to the operation manual. Cut off power during service.

7. An operator is responsible for obeying safety rules on the machine or in the operation manual and operating properly and working safely.

8. Contact with our Sales Service Center if there are any questions for security or proper operation.

9. Forbid removing and damaging the warning labels.

TASK 4

Tell the NOTES to your partner.

1. Never attempt to start the machine before closing the protective covers and doors to the full stop position.

2. Always close the front door before pressing the cycle start button.

3. Never try to open the doors while the machine is running.

4. Make sure that chuck jaws are firmly fixed before starting spindle.

5. Before loading or unloading the workpieces, retract the turret so that your hand will not be injured by the cutting tool on the turret.

6. When clamping the workpieces in the chuck, take care so that your fingers will not be caught by the chuck jaws.

7. Hold the workpieces in the chuck securely.

8. Never attempt to touch the workpieces or remove chips while the spindle is rotating.

9. The spindle does not stop immediately even when the spindle stop button is pressed. Never attempt to stop the spindle rotating with bare hands or tools.

10. Don't attempt hazardous operation.

Read, translate and fill in the blanks.

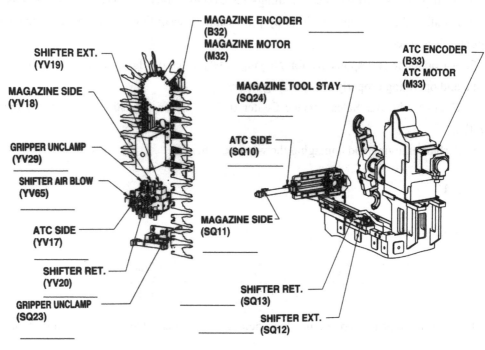

SHIFTER EXT.
(YV19)

MAGAZINE SIDE
(YV18)

GRIPPER UNCLAMP
(YV29)

SHIFTER AIR BLOW
(YV65)

ATC SIDE
(YV17)

SHIFTER RET.
(YV20)

GRIPPER UNCLAMP
(SQ23)

MAGAZINE ENCODER
(B32)
MAGAZINE MOTOR
(M32)

MAGAZINE TOOL STAY
(SQ24)

ATC SIDE
(SQ10)

MAGAZINE SIDE
(SQ11)

SHIFTER RET.
(SQ13)

SHIFTER EXT.
(SQ12)

ATC ENCODER
(B33)
ATC MOTOR
(M33)

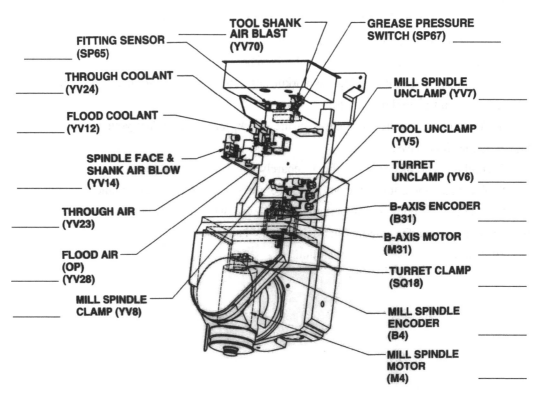

FITTING SENSOR
(SP65)

THROUGH COOLANT
(YV24)

FLOOD COOLANT
(YV12)

SPINDLE FACE &
SHANK AIR BLOW
(YV14)

THROUGH AIR
(YV23)

FLOOD AIR
(OP)
(YV28)

MILL SPINDLE
CLAMP (YV8)

TOOL SHANK
AIR BLAST
(YV70)

GREASE PRESSURE
SWITCH (SP67)

MILL SPINDLE
UNCLAMP (YV7)

TOOL UNCLAMP
(YV5)

TURRET
UNCLAMP (YV6)

B-AXIS ENCODER
(B31)

B-AXIS MOTOR
(M31)

TURRET CLAMP
(SQ18)

MILL SPINDLE
ENCODER
(B4)

MILL SPINDLE
MOTOR
(M4)

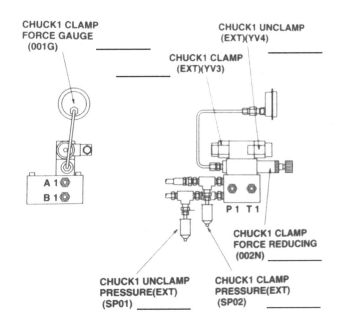

CHUCK1 CLAMP
FORCE GAUGE
(001G)

CHUCK1 UNCLAMP
(EXT)(YV4)

CHUCK1 CLAMP
(EXT)(YV3)

A 1

B 1

P 1　T 1

CHUCK1 CLAMP
FORCE REDUCING
(002N)

CHUCK1 UNCLAMP
PRESSURE(EXT)
(SP01)

CHUCK1 CLAMP
PRESSURE(EXT)
(SP02)

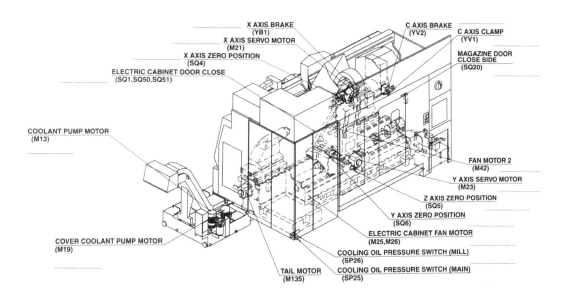

X AXIS BRAKE
(YB1)

X AXIS SERVO MOTOR
(M21)

X AXIS ZERO POSITION
(SQ4)

ELECTRIC CABINET DOOR CLOSE
(SQ1,SQ50,SQ51)

C AXIS BRAKE
(YV2)

C AXIS CLAMP
(YV1)

MAGAZINE DOOR
CLOSE SIDE
(SQ20)

COOLANT PUMP MOTOR
(M13)

FAN MOTOR 2
(M42)

Y AXIS SERVO MOTOR
(M23)

Z AXIS ZERO POSITION
(SQ5)

Y AXIS ZERO POSITION
(SQ6)

ELECTRIC CABINET FAN MOTOR
(M25,M26)

COOLING OIL PRESSURE SWITCH (MILL)
(SP26)

COOLING OIL PRESSURE SWITCH (MAIN)
(SP25)

COVER COOLANT PUMP MOTOR
(M19)

TAIL MOTOR
(M135)

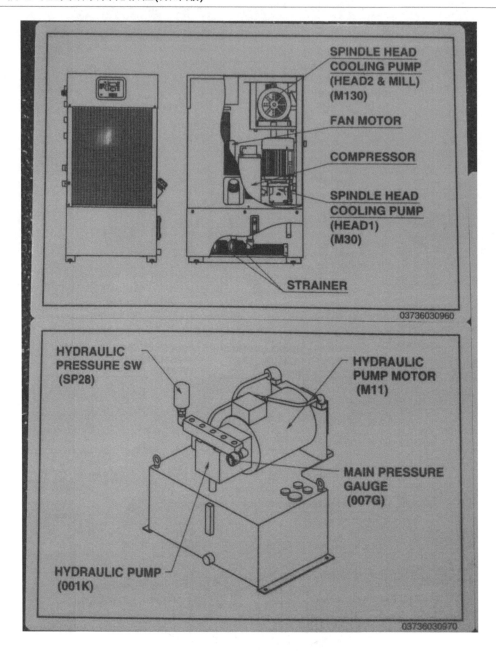

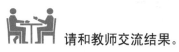 请和教师交流结果。

教师签名：_____

3.9   NC Operation

Task Content: Before working on the machine, Li Jun wants to learn how to produce a part on the computer by simulation. Please help him.

Translate the words or phrases in the word box into Chinese.

position	program	offset	shift
_____	_____	_____	_____
cancel	input	insert	delete
_____	_____	_____	_____
alter	graph	reset	end of block
_____	_____	_____	_____
message	system	page up	
_____	_____	_____	

Reading & Writing

There are some keys on the operating panel, shown as below.

1. **POS** POS means POSITION. Pressing this key, you can know the absolute position or relative position of the tool.

2. **PROG** PROG means PROGRAM. Pressing this key, you can begin programming.

3. **OFFSET SETTING** Pressing this key, you can set the offset value of the tool, either geometry or wear.

4. **CAN** Pressing this key, you can cancel the character before the cursor.

5. **INPUT** Pressing the key, you can input a program.

6. **DELETE** It is used to delete a word or a code.

7. **INSERT** Pressing this key, you can insert characters, one by one, or some at the same time.

8. **ALTER** Pressing this key, you can replace the word with another one.

9. ![key] Pressing this key, you can check whether the toolpath is true or false.

10. ![key] When pressed, the cursor returns to the beginning of the program. It can also be used to cancel the alarm messages. And when in Automatic mode, it can be used to stop the movement of the workpieces and tools, as well as coolants and programs.

11. ![key] ![key] Like that on the keyboard of a computer, these two keys are used to turn the page up and down.

12. ![key] It means the end of block.

Translate & Match

POS	POSITION
CAN	PROGRAM
PgUp	CANCEL
PROG	PAGE UP
DEL	INSERT
INS	DELETE
PgDn	PAGE DOWN

The following is the control panel of an NC machine. Please introduce it to your partner.

_____

_____

_____

_____

_____

_____

_____

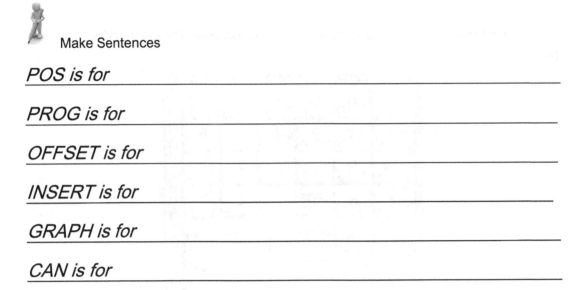

Make Sentences

POS is for _____

PROG is for _____

OFFSET is for _____

INSERT is for _____

GRAPH is for _____

CAN is for _____

ALTER is for _____

RESET is for _____

PAGE UP is for _____

PAGE DOWN is for _____

请和教师交流结果。

教师签名: _____

3.10　NC Programming

Tast Content: Li Jun now is studying the NC programming of a CNC machining center based on FANUC 0i series which reached school yesterday. Please help him.

Translate the word or phrases in the word box into Chinese.

coordinate system	unit	inch
_____	_____	_____
linear interpolation	code	METRIC
_____	_____	_____
program word	cutter compensation	cutter offset positive
_____	_____	_____
negative programming	fixed cycle	absolute position programming
_____	_____	_____
feed per minute	feed per revolution	tool number
_____	_____	_____
tool change	subprogram	coolant
_____	_____	_____

Reading & Writing

The control system of the machine is based on FANUC Oi series. There are different control functions. They are programmed through program words(codes).

1. **G-code** The G-code is also called preparatory code or word. G-codes are normally used as shown below.

code	function	code	function
G00	Rapid traverse	G41	Cutter compensation-left
G01	Linear interpolation	G42	Cutter compensation-right
G02	Circular interpolation, CW	G43	Cutter offset positive
G03	Circular interpolation, CCW	G44	Cutter offset negative
G04	Dwell	G80	Fixed cycle, cancel

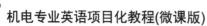

continue

code	function	code	function
G17	X-Y plane	G81-89	Fixed cycles
G18	Z-X plane	G90	Absolute dimension program
G19	Y-Z plane	G91	Incremental dimension
G20	Inch format	G54	Set the workpiece origin
G21	Metric format	G96	Constant surface speed control
G28	Return to reference point	G97	Constant spindle speed control
G32	Thread cutting	G98	Feed per minute
G40	Cutting compensation-cancel	G99	Feed per revolution

2. **F-Code** The F-Code specifies the feed speed of the tool motion. It is the relative speed between the cutting tool and the workpiece. It is typically specified in in./min(inch format) or mm/min(metric format). Ex. F100 means feed of 100mm/min.

3. **S-Code** The S-Code is the cutting-speed code. It is programmed in rpm. The S-Code does not turn on the spindle. The spindle is turned on by an M-code. To specify a 1000-rpm spindle speed, the program block is: s1000 m03.

4. **T-Code** The T-code is used to specify the tool number and the tool parameter. Ex, T0101

5. **R-Code** The R-code is used for cycle parameter to specify the safe height.

6. **M-Code** The M-code is called the miscellaneous word and is used to control miscellaneous functions of the machine, including turn the spindle on/off, start/stop the machine, turn on/off the coolant, change the tool, and so on. M-code is normally used as shown below.

code	function	code	function
M00	Program stop	M08	Mist coolant on
M01	Optional stop	M09	Coolant off
M02	End of program	M10	Chuck close
M03	Spindle CW	M11	Chuck open
M04	Spindle CCW	M30	End of tape
M05	Spindle stop	M98	Calling of subprogram
M06	Tool change	M99	End of subprogram (return to main program)

Reading & Match

G00	Rapid traverse _____
G01	Circular interpolation, CCW _____
G02	Linear interpolation _____
G03	Circular interpolation, CW_____
G04	Dwell _____
G32	X-Y plane _____
G41	Thread cutting _____
G54	Set the workpiece origin _____
G17	Cutter compensation-left _____

M00	Optional stop _____
M01	End of program _____
M02	Spindle stop_____
M03	End of tape_____
M05	Calling of subprogram_____
M06	Tool change _____
M30	Program stop _____
M98	Spindle CW_____
M99	End of subprogram _____

Interpret the program to your partner, and fill in the blank.

O0006;   _____

T0101;   _____

S1000 M03;   _____

G00 X52 Z10 M08;   _____

G01 X39.7 Z10 F50;　　　　_____

G32 X39.7 Z-20 F1.5;　　　_____

G01 X52 Z-20;　　　　　　_____

G00 X52 Z10 M09;　　　　　_____

M05;　　　　　　　　　　_____

M30;　　　　　　　　　　_____

 请和教师交流结果。

教师签名：_____

 微课资源

扫一扫：获取相关微课视频。

3.1 3D Printing-1-1

3.1 3D Printing-1-2

3.1 3D Printing-1-3

3.1 3D Printing-1-4

3.2 Engineering Drawing-1-1

3.2 Engineering Drawing-1-2

3.2 Engineering Drawing-1-3

3.2 Engineering Drawing-1-4

3.2 Engineering Drawing-1-5

3.3 Tolerances and Fits-1-1

3.3 Tolerances and Fits-1-2

3.3 Tolerances and Fits-1-3

3.3 Tolerances and

3.3 Tolerances and

3.4 Mechanical

3.4 Mechanical

Fits-1-4

3.4 Mechanical
Design-1-3

3.4 Mechanical
Design-2-1

3.6 Introduction of
AutoCAD(1)-1-2

3.7 Introduction of
AutoCAD(2)-2-1

Fits-1-5

3.4 Mechanical
Design-1-4

3.4 Mechanical
Design-2-2
3.6 Introduction of
AutoCAD(1)-1-3
3.7 Introduction of
AutoCAD(2)-2-2

Design-1-1

3.4 Mechanical
Design-1-5

3.4 Mechanical
Design-2-3
3.6 Introduction of
AutoCAD(1)-1-4
3.7 Introduction of
AutoCAD(2)-2-3

Design-1-2

3.4 Mechanical
Design-1-6

3.6 Introduction of
AutoCAD(1)-1-1

3.6 Introduction of
AutoCAD(1)-1-5
3.7 Introduction of
AutoCAD(2)-2-4

3.7 Introduction of
AutoCAD(2)-2-5